A POCKET GUIDE TO

NEWGRANGE

AND THE

BOYNE VALLEY

Gill Books, Hume Avenue, Park West, Dublin 12

www.gillbooks.ie

Gill Books is an imprint of M.H. Gill & Co.

Copyright © Teapot Press Ltd 2021

ISBN: 978-0-7171-8990-8

This book was created and produced by Teapot Press Ltd

Text by Fiona Biggs with Richard Killeen
Designed by Tony Potter & Becca Wildman
Picture research by Joe Potter, Freya Storch & Tony Potter

Printed in EU

This book is typeset in Garamond & Dax

5 4 3 2 1

A POCKET GUIDE TO

NEWGRANGE

AND THE

BOYNE
VALLEY

FIONA BIGGS

with

RICHARD
KILLEEN

Gill Books

Contents

Introduction

In 1849 Sir William Wilde, father of Oscar, wrote that the history of Ireland could be traced through its monuments. This is certainly true of the Boyne Valley region, where the monuments of Newgrange, Knowth and Dowth are older than the Pyramids, older than Stonehenge and older than the Celtic civilisation of Ireland. When newly built, these megalithic tombs, known to the Celts as Brú na Bóinne (the Mansion, or Palace, of the Boyne) and built in close proximity to each other in the Boyne Valley, would have dominated the rich, undulating landscape of this fertile part of eastern Ireland.

Brú na Bóinne is one of the world's most important complexes of Neolithic passage tombs. Newgrange, the largest, was built first, more than 5,000 years ago, around 3200 BCE (according to carbon dating of material excavated from the mound), by the Neolithic people who had settled in the Boyne Valley. The recent restoration of Newgrange allows today's visitors to the site to appreciate the awe that it must have inspired in those who built it.

An aerial view of the Bend of the Boyne.

All three burial sites were built on a broad ridge overlooking a stunningly beautiful stretch of the River Boyne, the Bend of the Boyne, as it enters a series of loops on its 112-kilometre journey from Carbery, County Kildare, to the sea at Drogheda, County Louth. The River Mattock encloses the area to the north, forming an 'island' that has made it an attractive place to settle for millennia.

The first stretch of the river is calm and tranquil, as it makes its way through the rich lands of County Meath, but as it approaches the sea it flows faster and its banks rise high above it. It flows through Tara, historic seat of the high kings of Ireland, which reached its zenith as a centre of power in the centuries before Christianity came to Ireland. Tara was finally abandoned in 1022 by High King Murchadha Ó Maeilsheachlainn. Fittingly, it was the location of Daniel O'Connell's biggest 'Monster Meeting' during the campaign for the repeal of the Act of Union.

Over the centuries the burial mounds gradually disappeared into the undergrowth, forgotten. However, their location, the Boyne Valley, has played an important role in Irish history and mythology. The setting for some of the country's most vivid Celtic legends, the Boyne Valley was also significant when Christianity came to Ireland in the fifth century; at that time this area was included in the kingdom of Brega (the kings of Brega referred to themselves as Rí Cnogba, Kings of Knowth). When St Patrick returned to Ireland he is believed to have landed at the mouth of the Boyne at Drogheda, and there is speculation that his famous Paschal fire, reputed to have been lit on the Hill of Slane, may in fact have been located at Knowth. The legend of St Patrick and the shamrock is also centred on the Hill of Slane. It is there that the saint is reputed to have plucked a shamrock to explain the Trinity to Christian converts, and the whole hill was covered in shamrock afterwards.

The arrival of Christianity brought large monastic foundations and these, in turn, became repositories of great wealth, ripe for plunder by Viking and native Irish invaders. Some of these invasion forces were large – in 837 at least 120 Viking ships were recorded on the Rivers Liffey and Boyne.

In the middle of the 12th century Christianity began to have an effect on the local landscape, especially with the establishment of the large Cistercian foundation of Mellifont (see page 176), near Drogheda, County Louth, just north of the Boyne Valley. The Cistercians would eventually control much of the land in Brú na Bóinne, where they founded Bective Abbey (see page 228). The monks were enthusiastic agriculturalists and were known for their creation of farms, or 'granges', on this outlying area of their extensive landholdings, and this gave Newgrange its name. By the late 12th century, when the Anglo-Normans arrived in Ireland, Knowth was in the middle of a Cistercian farm. The Cistercians were not alone in bringing Christian influence to the Boyne Valley. In 1180, the Augustinian order established a priory at Duleek (see page 130), also near Drogheda, on the site of an earlier abbey where King Brian Ború lay in state after his death at the Battle of Clontarf in 1014, the battle in which the Vikings suffered heavy defeat. At the end of the 12th century the Anglo-Normans built a church dedicated to St David at Dowth (it has since been replaced by a 14 – 15th-century structure) and later gifted it to the Augustinians.

Newgrange megalithic mound and the Boyne Valley from the top of the mound at Knowth.

A relentless campaign by the native Irish against the Anglo-Norman settlers led to the creation in 1488/89 of the Pale, a large area centred on Dublin and including parts of Louth, Meath and Kildare, extending from Castle Roche in the north (about 10 kilometres to the north-west of Dundalk) to Carrickmines Castle in the south, on the border between Counties Dublin and Wicklow.

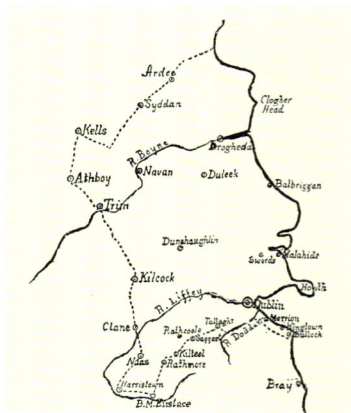

The Pale according to the statute of 1488.

Its northern and western borders were broadly defined by the course of the River Boyne. The Pale was protected by fortified ditches and ramparts, augmented with defensive tower houses, which shielded the resident population from attack. Brú na Bóinne was located in the middle of the Pale, with two fortified tower houses at Dowth.

With the Protestant Reformation in the 16th century came the dissolution of the monasteries and the establishment of a military administration at Dublin Castle. Although the confiscation of church lands was not as widespread within the Pale as in other areas of Ireland, and some Catholic landowners did manage to retain their estates, the Mellifont Abbey lands, including those at Knowth and Newgrange, were seized by the Crown and granted to the Protestant Moore family.

The 17th century brought more unrest to the Boyne Valley area. Trim, Kells and Drogheda figured in the 1641 Rebellion, which pitched Catholics against Protestants in a revolt against the suppression of Catholicism and the policy that 'planted' Protestants (largely in Ulster). The Catholic Confederacy, formed in the wake of the rebellion to promote the Catholic cause, was annihilated by Oliver Cromwell. In 1649, fresh from his victory in the English Civil War, the Lord Protector invaded Ireland. His campaign is infamous for a bloody siege at Drogheda (see page 138), which resulted in the slaughter of the entire garrison and many of the town's inhabitants. After a short but murderous campaign, Cromwell returned to England in 1650, having established Parliamentary control over Ireland.

The Boyne Valley is also notorious as the decisive battleground in the 17th-century power struggle between the Protestant William of Orange, who had been crowned King of England, Scotland and Ireland in 1689, and the Catholic King James II, who had been crowned in 1685, but was deposed by William in 1688, after the so-called Glorious Revolution. The Battle of the Boyne (see page 144), which laid the ground for centuries of unrest in Ireland, was fought on 1 July 1690, and involved the deployment of the largest number of troops (a total of 61,000 men) on an Irish battlefield. James was supported in his fight for the British throne by his Irish co-religionists, generally referred to as Jacobites. After a short battle, with surprisingly few casualties, given the numbers of troops involved on both sides, James fled the field, conceding victory to William. By the end of the year all Jacobite resistance in Ireland had been dealt with effectively and William's reign was secure. In 1691 his parliament introduced the Penal Laws, which limited the activities of Catholics in every sphere and continued until the 19th century. The Anglican Church became the established church in Ireland, although Catholicism, despite being outlawed, could not be suppressed.

In the 18th century, after the Williamite Wars, there was a period of prosperity and stability, and a general

Battle of the Boyne, 1 July 1690, between James II and William III. Painting by Jan van Huchtenburg.

William III of Orange at the Battle of the Boyne, where he defeated James II during the Williamite Wars. The battle took place near the town of Drogheda.

liberalisation of trade between Ireland and England led to the increased production and milling of grain and flax (for the burgeoning linen industry) in Brú na Bóinne. By the end of the century it was reported that there was a flax mill in every parish in the area. A new system of landholding brought about the creation of farms and demesnes with estates. The major landowning families in the area were the Nettervilles of Dowth and the Campbells of Newgrange and Knowth. Communications were improved with the construction of roads and the canalisation of the River Boyne to facilitate the transport of grain to Drogheda and to encourage trade with the capital.

Men on the Boyne Canal, early 1920s.

Having been in the eye of the storm during the rebellions and wars of the 17th century, the Boyne Valley area, through trade and good communications, had achieved considerable prosperity by the 19th century, and the area was not as badly affected during the Great Famine of the mid-19th century as were other parts of Ireland. It is one of the few areas of Ireland where there is a stock of labourers' housing – many of the original dwellings were replaced in the post-Famine period with solid, stone-built cottages.

In the late 1930s and early 1940s Irish-speaking families were moved from the West of Ireland to different locations in County Meath, including the Bend of the Boyne, necessitating the creation of small farms and the building of cottages. At this time, a number of 'pillbox' gun emplacements were erected in the Boyne estuary, a defence against the possibility of a British invasion. From the 1950s onwards, the modernisation of farming meant that the small cottages and patchwork landscape were absorbed into much larger holdings as intensive agriculture became the norm.

The success and prosperity of Brú na Bóinne in its Celtic and modern heydays – as evidenced by the valuable artifacts that have been discovered, the flourishing agriculture and productive linen industry – important though they are, pale into insignificance when compared with the engineering feats that produced the great Neolithic monuments that were constructed there. Newgrange, Knowth and Dowth are windows onto a civilisation that was shrouded in mystery for millennia.

In 1993 Brú na Bóinne was listed by UNESCO as a World Heritage Site, one of three on the island of Ireland, an accolade that legally obliges the state to protect the area to the highest international standards.

Standing stones at Newgrange, part of the stone circle surrounding the 5,200-year-old Neolithic passage grave.

CHAPTER 1 Who Built Newgrange?

In around 4000 BCE, less than a millennium before the construction of Newgrange, the Neolithic ancestors of the builders of the tomb may have come to Ireland from Scotland via the Antrim coast, or from the Iberian peninsula.

They travelled in sturdy rowing or sailing boats, made of animal hide stretched over a wooden frame. These crafts were not dissimilar to the traditional currach, still in use today. They would have been several metres in length, large enough to accommodate a family and their animals, together with the necessary supplies of food and water to sustain the occupants of the boats during the potentially treacherous ocean voyage.

It is the River Boyne that is key to the existence of Newgrange, Knowth and Dowth – its clear, fast-flowing waters were not only a direct source of food and water, but also a means of irrigating farmland, which was crucial to the settlements that came to populate that area of the Boyne Valley in the last era of the Stone Age, the Neolithic. Proximity to the river also provided the only possibility for travelling over longer distances in an era before the invention of the wheel.

In Europe, the Stone Age, named for the material used to make tools, was the period that followed immediately on the last Ice Age, predating the Bronze Age and the Iron Age. It is such a long stretch of time that it has been divided into three periods: Palaeolithic, Mesolithic and Neolithic. Ireland, where the glaciers receded later than in the rest of Europe, was still covered in ice for most of the Paleolithic, so the Irish Stone Age lasted from around 8000–2500 BCE, encompassing the Mesolithic and Neolithic eras. In the Mesolithic (the Middle Stone Age, from around 8000–6000 BCE) most people were nomadic, living as hunter-gatherers in temporary wooden dwellings. They foraged for berries and nuts, hunted birds and wild boar, and fished in the sea and rivers. Their weapons were flint-tipped wooden spears or harpoons and arrows. The Neolithic (the New Stone Age or Late Stone Age) lasted from around 6000–3000 BCE.

The practice of farming had gradually spread throughout Europe from the Middle East and it is one of the outstanding characteristics of the Neolithic. There is some evidence that the climate in Britain and Ireland became warmer and dryer around 4000 BCE, creating optimum conditions for agriculture. The Boyne Valley was an extensively wooded and very fertile area, so it was a draw for people moving from the nomadic hunter-gatherer lifestyle of the Mesolithic to a settled, agricultural life.

The introduction of farming changed everything. The people who came to Ireland at this time (around 4000 BCE) formed settled populations that had, for the first time, a fixed dwelling place and depended for their food on grain crops, domestic livestock (cattle, sheep and goats, but no chickens) and some hunting and foraging. They didn't come as invaders or conquerors, but may have lived alongside the native hunter-gatherer population, gradually absorbing it, until the whole population of the island depended on agriculture.

River Boyne, County Meath

We know nothing of the language, social structure, religious beliefs or customs of the Neolithic population that settled in the Boyne Valley, but we can speculate that while women would probably have participated equally in hunting and gathering activities in the Mesolithic, the Neolithic concentration on farming may have produced a division in the roles required for the successful maintenance of the group, with men occupying themselves almost exclusively with tillage and hunting. It is possible that it is at this point that women first became more engaged in domestic work, childcare, care of domestic animals and gathering plants for food.

In order to clear land for farming the settlers cut down or burned woodland, so that they could create areas for grazing and for planting. They tended to settle areas of higher ground, where the forest cover was sparser and therefore easier to clear. As there were no horses in Ireland at that time, ploughs were most likely pulled by cattle. Although the farmland the settlers created was fertile, overgrazing and erosion as a result of their rudimentary approach to agriculture eventually caused it to stagnate, providing the basis for some of the peat bogs that are a feature of the modern Irish landscape. The Céide Fields in County Mayo provide solid evidence of Neolithic agricultural practice and the fate of the land that was farmed in this era. This 500-square-metre piece of land was discovered in the 1930s under a peat bog at the edge of a cliff to the north-west of Ballycastle. When the site was excavated in the 1970s it was found to have been divided into fields, separated by stone walls – these are the oldest stone-walled fields in the world, and may have been typical of Neolithic farms.

The quern stone was indispensable to the Neolithic diet, depending as it did on grains. Not many grains could be grown in Ireland at the time, but an ancient type of wheat and barley had adapted to the Irish environment. Like all grains, they had to be ground in order to be digestible, and the most efficient method of doing this was to grind them between two stones. Domestic stone grinders of various kinds, dating back to long before the Neolithic, were used on every continent. Saddle querns have been excavated at several Neolithic sites – the lower stone is flat or slightly convex and the upper stone is cylindrical, somewhat like a rolling pin. The grain was placed on the lower stone and the upper stone was repeatedly rolled over it until the grains broke down into flour. A type of flatbread made with flour mixed with water was cooked over the open fire.

Pottery was introduced in the Neolithic – it was made using the simple coiling method, where a rope of clay was rolled and then coiled upwards, then shaped into a simple pot and smoothed down before firing in the open fire. Some pots were decorated before firing with designs etched on them with sharp-pointed sticks, but these vessels would have been primarily functional rather than decorative. Pots were indispensable to the Neolithic way of life – they were

essential for carrying and storing water and for keeping long-lasting foods such as the wheat and barley that was used throughout the colder seasons. They could also be filled with animal fat, which was ignited to provide light.

Because the introduction of pottery made it possible to store food for leaner seasons, people no longer had to band together in hunting groups all year round to roam the land in search of food. They were able to build permanent settlements in the middle of the farms that they created. Because they were settled, Neolithic people are believed to have lived in larger groups than had been possible in the Mesolithic.

Animals were important as a source of meat and skins. There is some evidence of textile weaving in the Neolithic (fibre degradation makes it difficult to tell what patterns, if any, were used by the weavers), but animal hides were necessary for warm clothing, bedding and foot coverings.

Neolithic pottery from Toorne, County Antrim, in the Ulster Museum.

Scrapers, most likely used for scraping hides, have been excavated at Neolithic sites, indicating that some clothing was made of leather. Textile disintegration means that there is no real way of knowing what fibres were used to sew these garments together – suggestions include strands woven from animal hair, human hair, flax and tree fibres, even grass.

What type of dwelling did these new settlers build? At the Brú na Bóinne Vistor Centre there is a reconstruction of the type of hut that they may have built as shelters. It is circular at the base and cone-shaped, somewhat resembling a teepee. The walls are made of wattle and daub (a technique that seems to have come into use at around this time, whereby a finely woven wooden lattice was 'plastered' with wet clay or animal dung), the roof is supported with wooden beams and the whole structure is thatched with reeds pulled from the river. Fires were brought indoors for the first time in the Neolithic and were made in a fire pit in the centre of the dwellings. A hole in the roof allowed the smoke to escape. The huts were grouped together in small clusters, but, although stockades were built to protect the livestock, there were no fortifications, indicating that there was no human threat to the settlers.

At the Ulster History Park in County Tyrone there is a reconstruction of a much larger, rectangular Neolithic dwelling, again finished in wattle and daub, but with a thatched pitched roof. It is thought that the larger dwellings may also have been used for food storage and housing domestic animals.

Although creating farms required the destruction of forests, the large Boyne Valley woodland would have been vital to the settlers in the area, and would have been a reason for them to choose this as the location in which to put down roots. Wood was used as a building material, as a source of heat and for making utensils, farming equipment and furniture. It was also an integral part of the settlers' weaponry, used for protection and in hunting.

In the Mesolithic, flint was the only material suitable for making weapons and sharp utensils. Large deposits of flint on the northern coast of Ireland were accessible from the Boyne Valley. However, the Neolithic settlers also used porcellanite, a dense type of rock that looks like unglazed porcelain, but is much harder than flint. It was particularly useful in the making of axes, numerous examples of which have been excavated. The only sources of porcellanite in

Ireland are at Cushendall on the Antrim Coast and on the offshore Antrim island of Rathlin. This would have made it particularly accessible to the Boyne Valley communities.

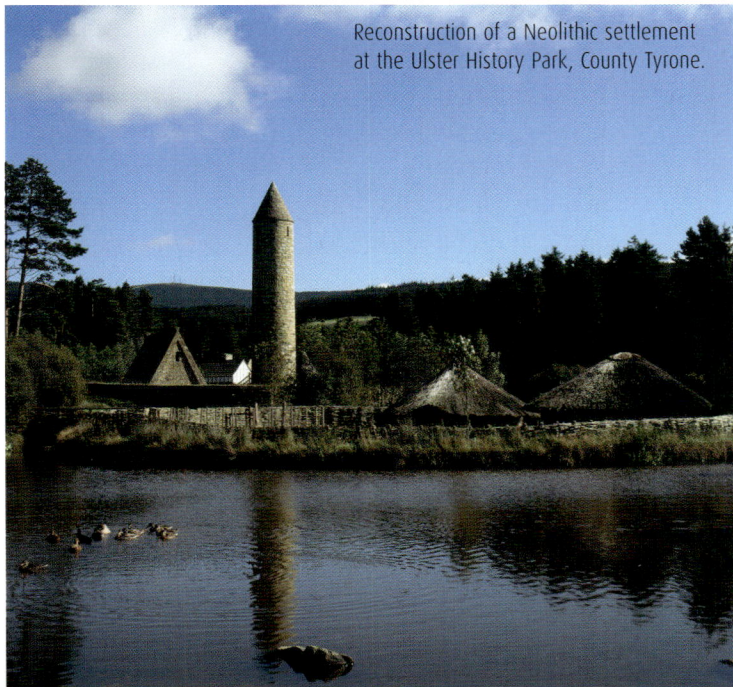

Reconstruction of a Neolithic settlement at the Ulster History Park, County Tyrone.

We can't see into the minds of the Neolithic settlers, so we can only speculate as to why they built Newgrange (and nearby Knowth and Dowth). With the introduction of farming, communities began to settle in one place. Settlement meant that people were buried at marked locations close to the community and it is believed that the passage tomb at Newgrange may have been part of a tradition of remembering or venerating the dead.

In the early Neolithic, the megalithic (mega = large; lith = stone) tradition of tomb-building had spread across Western Europe, from as far north as Sweden, as far east as Poland, and as far south as the southern coast of Portugal. The oldest passage tombs in Europe are located in Brittany, and they were built at least a thousand years before Newgrange, in around 4000 BCE. One of the earliest is the Cairn of Barnenez in Finistère, widely regarded as one of the oldest structures in the world that is still standing. Radiocarbon dating suggests that it was built during the first half of the fifth millennium BCE. Equally impressive is the cairn at Gavrinis, built around 3500 BCE on a small offshore island in the Breton gulf of Morbihan. Gavrinis has many points of resemblance with Newgrange, and, as there, the winter solstice sunrise shines into the inner chamber of the tomb.

Newgrange, one of more than 200 Irish passage tombs, was not the first to be built on the island – the complex at Carrowmore in County Sligo, which has similarities with several tomb complexes in south-west Wales, is just one example of an earlier one, dating from around 3800 BCE. However, before Newgrange was built, these tombs and tomb complexes were much simpler, allowing for reopening to add new remains, but with very restricted access to the interior of the tombs, so there would have been no possibility of a small gathering in the tombs' interiors as there is at Newgrange, Knowth and Dowth.

It is the possibility of small gatherings in the interior of the larger tombs that has given rise to speculation that they were built for religious reasons, to accommodate gatherings for the veneration of ancestors or the performance of funereal or other rituals.

Newgrange from the air.

Newgrange: interior of the tomb.

However, recent evidence supports the idea that Newgrange was used for royal burials. In June 2020, the results of a study surveying ancient Irish genomes was published. It found that the parents of a man buried there were brother and sister, or other first-degree relatives – Inca kings and Egyptian pharaohs were closely related in a similar way. This suggested to the researchers that the people who built the tomb lived in a hierarchical society with a ruling elite that was allowed to break normal taboos. Dr Daniel Bradley, a specialist in ancient DNA at Trinity

College Dublin, described the remains as those of 'an Irish Pharaoh'. Researchers involved in the project also found that the man was related to people who had been buried in other megalithic sites across Ireland, including Carrowkeel and Carrowmore in County Sligo. According to Dr Lara Cassidy of Trinity College, this is confirmation of 'a powerful extended kin-group, who had access to elite burial sites in many regions of the island for at least half a millennium'.

Others ascribe a religious significance to the monument. Did the Neolithic have a cult of the dead? Or perhaps, given the specific solar alignment of the tomb, their beliefs were based on astronomy? The activity of the sun, giver of life and warmth, would have been of paramount

importance in the lives of the Boyne Valley settlers. Other ancient monuments, such as Stonehenge, were aligned with solar, lunar or stellar events – Stonehenge, for example, is oriented towards the sunrise on the longest day of the year, Midsummer's Day, 21 June in the modern calendar. In Ireland alone, about 24 sites have a significant solar orientation, most of them towards the rising or setting solstitial sun. In the course of his excavation of Newgrange in the 1960s, Michael O'Kelly noted the orientation of Newgrange towards the sun at the winter solstice – the low rays of the winter sun enter the tomb on several days around 21 December, the shortest day of the year in the northern hemisphere. Five thousand years ago the light spectacle would have been enjoyed for longer, as the sun

A stylised representation of Newgrange at sunrise.

would have entered the chamber every morning for a full week on either side of the solstice, bringing with it the hope of rebirth at the darkest time of the year. It was an anticipation of the coming of spring and the fruitfulness of summer – something worth contemplating even now with joy and optimism, despite our separation from the seasonal rhythm of life, but for our Neolithic forebears it represented survival.

This alignment towards the sun at the winter solstice means that for the builders of Newgrange, the sun and its cycle must have had a profound significance, possibly even a religious one. Some believe that the orientation may have been an attempt to capture the light of the sun so that it could be used to lengthen the coming days. The attitude to the sun may have been born of simple necessity, for without it there is no life. Climate change during the previous century would have caused concern – perhaps the people of the Boyne Valley felt a need to honour the sun so that their continued existence could be guaranteed?

There were similar astronomical alignments at Knowth and Dowth (which also has a winter solstice orientation), giving rise to the idea that the mounds may have been constructed with the aim, at least in part, of marking the passage of time in some sort of calendrical way. O'Kelly was of the view that all three monuments, 'cathedrals of megalithic religion', had to be looked at in conjunction with each other. Taken together with the many other passage tombs in Ireland, Newgrange is convincing evidence of a 'cult of the dead', but this does not contradict the existence of a set of religious beliefs centred on the sun. If it was indeed a royal burial place, it makes sense that an elite site of interment would have been located in a structure of immense astronomical importance.

Tourists inside the mound at Newgrange.

Whatever the reasons for building Newgrange, the precision with which the alignment was made is indicative of highly developed skills of astronomy and engineering. Until the first cloudless winter solstice after the completion of the structure, the builders would have had no way of knowing if their calculations had been absolutely correct.

On 21 December 1967, O'Kelly became the first person in the modern world to witness the winter solstice sunrise in the central passage at Newgrange (see page 78). Since his discovery of the orientation of the Newgrange chamber, many people have tried to imagine the emotions of those standing in the chamber for the first time to witness the rising of the solstitial sun. Archaeologist Robert Hensey witnessed the sun entering the chamber at Newgrange in 2011 and wrote this account of his experience:

And then it happened. It began with pale ambient sunlight in the passage and chamber. A richer coloured line of light was visible along the edge of this band, like the hem of a dress or the measurements along the edge of a ruler … for a brief moment I managed to sink beneath my own expectations, the inevitable grasping after experience, to taste another perspective. I imagined Neolithic people waiting, really waiting, not entering the chamber minutes before the event as we had done, but remaining in the chamber for perhaps hours or days previously.

As the group watched, all intensely focused on minute changes in the light, it seemed that we became a single entity; our separate identities temporarily forgotten in the wonder of the moment. The deeper line of sunlight started to increase in width, slowly expanding into the wider band of pale ambient light. But this was no static artificial light; it changed continually; not only with its entry and its exit into the monument and its variation in width, but in its tone and colour. Tiny dust particles floated in the sun's light creating the illusion of activity mixed with stillness. It is easy to imagine how the light could have been perceived as animate in the past … Then, slowly, almost as mysteriously as it had entered, the light began to narrow and start to recede from the chamber. The event was over.

(Robert Hensey, *First Light: The Origins of Newgrange*, Oxbow Books, 2015)

CHAPTER 3 How was Newgrange Built?

Newgrange is a hugely impressive monument, even to modern eyes, and the effort and ingenuity required to build it would have been enormous. Unlike the builders of the pyramids in Egypt a thousand years later, the Boyne Valley people who built Newgrange left no written accounts of their lives. Because of the absence of any records other than the edifice they left behind, in some minds, perhaps, the builders of this stunning monument were an uncivilised people. However, it is clear that they had exceptional engineering and organisational skills and an advanced knowledge of architecture, astronomy and geology. The Neolithic inhabitants of the Boyne Valley must have been very sophisticated in order to conceive of and plan the building of the tomb, and sufficiently wealthy to have had the resources with which

One of the 12 standing stones at Newgrange.

Part of the revetment wall at Newgrange.

to execute their plans. It is believed that Newgrange is the oldest structure in the world that was built with a deliberate astronomical alignment.

When making any assessment of Newgrange, it has to be remembered that it was built without metal tools and before the invention of the wheel, so it would have involved a huge physical effort. The workforce would have been drawn from the general population. As this was a farming community, no labour would have been available during the busy agricultural seasons, so work would probably have proceeded slowly, with construction carried out for two or three months each year over a considerable number of years. Modern estimates of the length of time required for the construction range from a mere five years, to at least thirty years. It may even have lasted generations, so that the original builders of Newgrange would not have been alive to see its completion.

The first task would have been to choose a site. Newgrange was built on an elevated site with a commanding view of the countryside, but the gentle topography of the Boyne Valley means that the position of the monument lacks the elevation and vantage point of, for example, Carrowkeel in County Sligo, with its 360° views across the western counties of Ireland, or even Knowth, a few kilometres to the west of Newgrange, with a vista as far south as the Wicklow Mountains.

Once chosen, the site would have been cleared of trees, a large enough undertaking in this heavily wooded landscape. Only then would the careful calculations for the siting of the mound have begun.

For its architects, the most important aspect of the mound would have been its alignment. Before building work began they would have made careful calculations based on the position of the sun as it crossed the sky during the year. The solar observations must therefore have taken at least a year, but more probably several years, in order to ensure that the rising sun at the winter solstice would shine through the roof-box directly into the inner chamber.

Carrowkeel tombs. A Neolithic passage tomb on top of the Bricklieve Hills, County Sligo.

Newgrange: interior view from just inside the entrance as the winter sun rises.

The degree of precision required must have been daunting, given that they had no sophisticated instrumentation with which to make their calculations. Five thousand years ago the thin beam of light as the sun began to rise would have hit the stone at the back of the chamber in a precise spot; changes in the earth's orbit around the sun since then mean that it no longer does so, as the sun rises at a slightly different point on the horizon. The rays now hit the stone basin on the floor of the chamber and then move down the passage. Although they must originally have entered the chamber at the precise moment the sun appeared on the horizon, this now happens four or five minutes after the beginning of the sunrise.

Once the calculations had been made and the plan of the mound had been established, the gathering of the necessary building materials would have been organised and realised. This would have been an enormous task, especially as all the building materials had to be carried uphill to the site of the mound. Michael O'Kelly, who carried out extensive investigations of the mound in the 1960s and 1970s, thought that the workforce would have been divided into as many as six groups or gangs – one group would have sourced the large slabs from as far afield as the Dublin Mountains, Wicklow Mountains and Mourne Mountains; another would have set up the orthostats and other structural stones, and would eventually have placed the kerbstones; a third group would have gathered and deposited the stones for the cairn; another group would have caulked the seams with an earth and sand mortar and covered the mound in turves; the group responsible for the wooden aspects of the build would have cut trees and made the planks, rollers and pallets that were used to transport the building materials. The last group comprised the artists, the creators of the beautiful carvings on the stones.

Most of the main building blocks used were 'greywacke', a type of rough sandstone. Greywacke rocks would have been found on the riverbed of the Boyne, so they were easily accessible from the site. The rocks that were found were used as they were, without any shaping or dressing. Fewer than 500 were used in the building of the mound, but some of them were very large, each weighing at least a tonne. The largest rock is more than four metres in length, and weighs several tonnes. Transporting the rocks uphill

from the valley floor to the site, over a distance of one kilometre, must have presented many difficulties.

There is nothing to indicate how the rocks were transported up the hillside, so we can only speculate as to how this was achieved. They may have used teams of oxen or they may have pulled them themselves, tying ropes around the rocks and pulling them upwards along beds of logs, which would have provided traction. It is likely that they would have built earthen ramps to facilitate this log-rolling effort. While this method would have made transportation possible, it would not have made it easy. Once the rocks were at the site they would somehow have been levered into position.

Most of the Newgrange mound is composed of smaller, water-rolled stones, probably also collected from the river and then carried to the site by individual labourers, probably in baskets carried on their backs. This would have been exhausting work, and the task must have seemed endless.

The mound is supported at its base with 97 greywacke kerbstones, none of which weighs less than a tonne, forming a sort of retaining wall. Many of the kerbstones are beautifully carved with elaborate symbols and motifs. The most outstanding example of these decorated kerbstones is the iconic one at the entrance to the tomb. These kerbstones withstood the collapse of the small loose stones of the mound, which had been packed into place with sods of earth. Because the decoration on many of the kerbstones appears on the underside and back of the stones as well, it was suggested that many of them had been carved for another location or locations and are an early example of repurposing, but this was disproved by Michael O'Kelly, who excavated the site. If the decoration means anything, and it is likely that it does, nobody has been able to decipher it as yet. We can speculate that it is tied into the calendrical purpose of Newgrange and represents the seasons or some other method of marking the passage of time, but no one has been able to provide a definitive assessment of it.

Kerbstone carving, detail.

The construction of the roof of the chamber was an extraordinary piece of engineering by any standards. It was built with overlapping layers of flat stones topped off with a capstone, a method known as 'corbelling'. As each layer of stones was put in place, their downward pressure caused the previous layers to become more tightly locked together. It is likely that there was one team inside the mound, and another on top. The people working in the interior would have ensured that the slabs were correctly laid, and they then pick-dressed the edges to round them off. The joints were caulked with a sort of putty made from soil and burnt sand so that the roof would have been completely waterproof. The distance from the floor of the chamber to the capstone is six metres, so scaffolding must have been erected. In the larger, right-hand recess of the chamber there are two stone basins, one placed inside the other. The larger of these is too big to have been brought through the passage, so it must have been lowered into the chamber before the capstone was put in place.

The roof stones along the length of the passage had grooves cut into their upper surfaces to run off any rainwater that might otherwise have accumulated. The roof is so well engineered that although the pressure created

The Tuatha Dé Danann

The Celts believed that the Tuatha Dé Danann (the People of the Goddess Danu), or the Tuatha Dé (the People of the Gods) were a mythological race endowed with magical powers. They lived in the mounds at Brú na Bóinne and other such sites and, despite their supernatural abilities, engaged with humans on a daily basis. Much of the lore about the Tuatha Dé was written down during the Christian era, and some Christian luminaries, such as Brigid, found their way into the reconstructed Celtic pantheon. Eventually, the Tuatha Dé became synonymous with the Sídhe, or fairies, of later folklore – in the popular imagination, the great burial mounds of the Neolithic became fairy mounds. It is in these legends that the great mound at Newgrange gets the name Brú na Bóinne.

The Tuatha Dé Danann, as depicted in John Duncan's *Riders of the Sídhe* (1911).

The Dagda

The Dagda was a relatively benign being at the top of the Celtic hierarchy of gods. He was associated with the sun and his domain was Brú na Bóinne. He possessed several magical objects: a staff, one end of which was used for killing, while the other was used to bring forth life; a cauldron that was never empty; a harp that could control people's emotions and change the seasons at the Dagda's whim. For a god, he seems to have had a limited life span – he was killed in battle after a 70-year reign.

The Dagda had an affair with Bóann, the goddess of the River Boyne, and they had a son, Aengus Óg. When Aengus was away, the Dagda shared out his property between his other children. When Aengus returned he insisted that his father give him Newgrange. The Dagda refused outright, but eventually agreed that Aengus could stay there for a day and a night, forgetting that this would confer ownership of the mound on his son. He was thus tricked into relinquishing ownership of his palace.

Cúchulainn

Newgrange was said to be where Cúchulainn was conceived, and the Boyne Valley is the backdrop to many of his exploits. These feature in many of the old tales, not least the Táin, where it is recounted that the young hero singlehandedly rebuffed the army of the powerful Queen Maeve of Connacht, who had used magic and subterfuge to steal the Donn Bó Cuailgne, the Brown Bull of Cooley.

Fionn Mac Cumhaill

The larger-than-life hero of many of the Celtic myths and legends is Fionn Mac Cumhaill, leader of an elite group of warriors known as the Fianna. He has either a central or peripheral role in most of the Celtic legends. One of the earliest stories about Fionn is set on the banks of the River Boyne and provides an insight into the reason for his outsized reputation.

Fionn Mac Cumhaill meets his father's old retainers in the forests of Connacht; illustration by Stephen Reid.

Fionn Mac Cumhaill and the Salmon of Knowledge

When Fionn was born his father, who was the leader of the Fianna, sent him to be fostered by a wise woman who lived in a forest. One day, while the young boy was out hunting beside the River Boyne, he came across an old man, Finnéigeas, who had just caught an enormous salmon. He was about to cook the salmon, and Fionn offered to help him with it, by gathering firewood and lighting a fire.

While Fionn was setting the fire the old man told him that the fish he had just caught was the Salmon of Knowledge, and whoever ate it would have all the knowledge of the world – knowledge of things past and things yet to come. He warned the boy not to eat any of the fish, even to test it, while it was cooking.

Fionn assured the old man that he wouldn't taste the fish before he brought it to him, and the old man went to his cabin to have a sleep before his magical meal.

Fionn cleaned and prepared the fish and set it to cook over the fire. When he thought it was cooked he lifted it carefully from the fire, but a blister on the salmon's skin burst, spattering hot fat on Fionn's thumb – he put his thumb in his mouth to cool the burn and he immediately

Fionn Mac Cumhaill brings the Salmon of Knowledge to Finnéigas.

realised that he knew everything that there was to know in the world. He brought the fish to the old man, who knew as soon as he looked at Fionn that he had tasted the salmon.

He was sad rather than angry, and asked Fionn what his name was. When the boy told him, the old man said that in that case the prophecy had been fulfilled – someone called Fionn was destined to eat the Salmon of Knowledge. He told him to eat the whole fish so that he would receive every scrap of knowledge, and that if he ever needed to access that knowledge he just had to put his thumb in his mouth.

The Pursuit of Diarmuid and Gráinne

Fionn is usually the hero of the Celtic myths, but one important story, which features Brú na Bóinne, characterises him as spiteful and vindictive. Diarmuid and Gráinne's story is in many ways timeless, involving a beautiful young woman, an elderly suitor and a young and virile lover.

Gráinne, daughter of High King Cormac Mac Airt, was somewhat unwillingly betrothed to the ageing Fionn Mac Cumhaill. At a great feast held at Tara to celebrate the betrothal, Gráinne encountered Fionn's favourite warrior, Diarmuid, for the first time, and fell instantly in love. Diarmuid didn't reciprocate and, in any case, was loyal to Fionn, but Gráinne slipped a magic potion into his drink to make him fall in love with her. They eloped together and Fionn set off with the Fianna in hot pursuit. Aengus, son of the Dagda, was Diarmuid's foster father, and he intervened to protect his foster son and his wife. He managed to pacify Fionn to the extent that he declared a truce, but the insult rankled. Several years passed, and one day Diarmuid was injured while out hunting wild boar. Fionn had magical healing powers that he could have used to save him, but he refused and Diarmuid died. Heartbroken, Aengus brought Diarmuid's body to Newgrange for burial.

Contemporary illustration of the story of Diarmuid and Gráinne.

Cormac Mac Airt

Cormac is one of those figures who straddles myth and fact in ancient Irish lore. Also known as Cormac Ulfadha (Long Beard), he was a legendary high king with a palace at the Hill of Tara and a beautiful daughter, Gráinne. It is thought that such a person may have reigned in Ireland in the first half of the third century AD. The Book of Leinster records that Cormac hosted both Romans and Gauls. Although he died in around 266 AD, he features in stories of St Patrick that are located in the Boyne Valley, probably in an effort to bolster the message of Christianity preached by Patrick.

A Christianity-imbued version of the Cormac legend relates how he made a deathbed request not to be buried at Newgrange with the pagan kings, but at Ros na Rí, facing the Christian dawn. His druids ignored his wishes and set out for Newgrange with his remains. In order to get to Newgrange from Tara, they had to cross the River Boyne. They tried three times to cross the swollen river but were repulsed. On the third attempt a huge wave broke the banks of the river, sweeping Cormac's body away, eventually depositing it at Ros na Rí, where he was buried.

King Cormac
Ulfadha meets
Eithne Ollamhdha,
and makes her
his queen.

The poet W. B. Yeats, a leading light of the Celtic Revival movement, wrote on themes that bridge the gap between the ancient Celtic world and the superstitions that grew up around the 'fairy mounds'. 'The Song of Wandering Aengus', first published in 1897, is based on a story about the god's wanderings in pursuit of the woman of his dreams. The last two lines of this poem are well known.

I went out to the hazel wood,
Because a fire was in my head,
And cut and peeled a hazel wand,
And hooked a berry to a thread;
And when white moths were on the wing,
And moth-like stars were flickering out,
I dropped the berry in a stream
And caught a little silver trout.
When I had laid it on the floor
I went to blow the fire a-flame,
But something rustled on the floor,
And someone called me by my name:
It had become a glimmering girl
With apple blossom in her hair
Who called me by my name and ran
And faded through the brightening air.
Though I am old with wandering

Through hollow lands and hilly lands,
I will find out where she has gone,
And kiss her lips and take her hands;
And walk among long dappled grass,
And pluck till time and times are done,
The silver apples of the moon,
The golden apples of the sun.

An earlier poem, 'The Stolen Child', published in 1889 in *The Wanderings of Oisín and Other Poems*, taps into the people's fear of the power of the fairies and their 'lios', or fairy mound. It tells the story of a child entranced by the fairy people, and contains these wonderful lines, repeated as a chorus at the end of each verse:

Come away, O human child!
To the waters and the wild
With a faery, hand in hand.
For the world's more full of weeping than you can understand.

The old stories that passed down from generation to generation had a powerful impact on people's imaginations. Far from exciting investigative curiosity in the locality, Newgrange and the other mounds in the Brú na Bóinne were, for several centuries, regarded as fairy mounds, places to be avoided.

In 1699, after the Williamite Wars, the lands of Mellifont Abbey (see page 176) at Newgrange were granted to Charles Campbell on a 99-year lease. Having discovered that there was stone beneath the grass of a large mound on his land, he set workers to quarrying it out of the side of the mound. In the course of their work, they discovered what seemed to them to be a cave, with a decorated stone at its entrance. The Welsh antiquary Edward Lhwyd was visiting Ireland at that time; hearing of the discovery, he went to Newgrange and took detailed notes, describing the mound as a cave with a 'rudely carved' stone lying across the entrance. He found a Roman coin at the site (later Roman finds included finger rings, a fragment of a gold torc and a number of gold coins) and concluded that the site was older than the Viking invasions, probably an ancient Irish place of sacrifice or burial. He also noted the presence of a standing stone on top of the mound – a contemporary drawing of the mound at Newgrange by John Anstis, who made extensive records of ancient monuments in Britain and Ireland, shows this stone. In 1726, Thomas Molyneux published his findings based on an earlier visit – he thought the remains were from the Viking era (other investigators speculated that it was built by the Phoenicians or

Opposite:
Mellifont Abbey, the first Cistercian foundation in Ireland.

Part of a cache of Roman jewellery found at Newgrange.

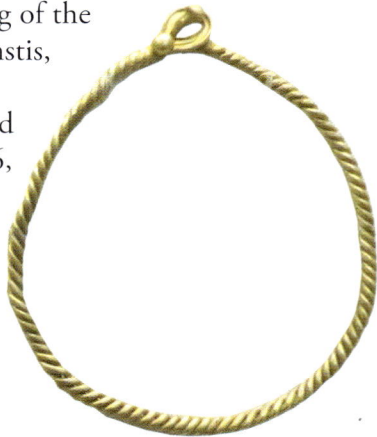

ancient Egyptians). In 1769 Thomas Pownall sent a letter to the Society of Antiquaries in London asserting that Newgrange was the remains of a much larger construction. He originated the idea that the decorated stones there had been imported from an older structure.

These early investigators were followed by more in the 18th and 19th centuries, all adducing different theories as to the purpose of the mound, and most of them unwilling to accept that its construction predated the Celtic era. In an 1833 issue of the *Dublin Penny Journal* antiquarian George Petrie was the first person to assert that Newgrange had been built not by Danes or Normans, but by Irish people. In 1849, in *The Beauties of the Boyne and Blackwater*, William Wilde identified the location of Brúgh na Bóinne, and detailed 17 burial mounds within that area. He also wrote up the site at Knowth, providing the first illustration of the mound. It was at this time that the decorated Newgrange entrance stone and roof-box (then believed to be a lintel) were unearthed.

1847 saw the beginning of the only investigation of Dowth, led by an engineer, R. H. Frith. He discovered the central tomb and its annexe and a souterrain, but found no Neolithic remains.

When Frith stopped working on the mound two years later he left it unprotected and it was subsequently plundered locally for its stone.

William Wilde's engraving of the entrance stone.

By this time, the Brú na Bóinne monuments had begun to attract the interest of members of the public. Unsupervised visitors inevitably caused damage and some even left a permanent mark in the form of graffiti (some can be seen in the passage at Newgrange). The site was also suffering from the presence of farm animals. A 19th-century folly a few metres behind the mound at Newgrange was built using stone taken from the mound. In 1882 the Ancient Monuments Protection Act placed Newgrange, Knowth and Dowth in the care of the state, and in 1890 some repair and conservation work was carried out by the Board of Public Works. In 1916, George Coffey, Keeper of Irish Antiquities in the National Museum of Ireland, published *New Grange and Other Incised Tumuli in Ireland*, which provided a comprehensive listing of the monuments of the Brú na Bóinne and catalogued the decorated stones. Coffey's labelling system is still in use today.

In the 1920s, some investigations were carried out at Newgrange by R. A. S. Macalister, Harold Leask and Robert Lloyd Praeger. They exposed 54 of the kerbstones, which Macalister believed were gravestones taken from another site. The tenant farmer on the Newgrange land objected to the work and it was stopped, and in 1941

Victorian folly built at Newgrange using stones taken from the monument.

Macalister began investigating Knowth, revealing half of the kerbstones and a burial chamber in one of the 17 satellite tombs.

In the period after the Second World War, and especially after 1954, when electric light was installed in the passage of the tomb, increasing numbers of visitors came to the three sites, and the possibility of serious damage led to the purchase of three hectares of land at Brú na Bóinne by Bord Fáilte. It was handed over to the Commissioners of Public Works, and Ireland's leading archaeologist,

Kerbstone 52 at Newgrange.

Professor Michael J. O'Kelly, was asked to carry out exploratory work in advance of a restoration project.

Summer 1962 saw the initiation of the excavations – they continued for four months every summer until 1975. Looking at the smooth grassy surface of the mound today, it is difficult to imagine that in 1962 it was covered in scrub and dotted with trees – all of these had to be cleared before work could begin. The most important task after this was to examine the bank of stones (the 'cairn slip') around the entrance and around Kerbstone 52, on the north side of the mound, diametrically opposite the entrance stone. Kerbstone 52 had fallen onto its face, so its decoration was revealed only when it was righted. The elaborate nature of the decoration on that stone led O'Kelly to believe that there might be a second passage and chamber behind it (as is the case at Knowth), but nobody has ever managed to find one.

Newgrange tumulus chamber cross-section
drawing, based on an original sketch.

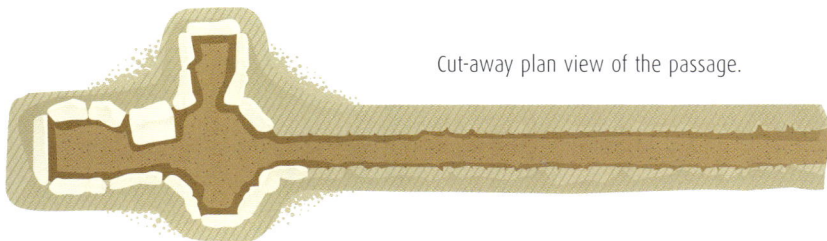

Cut-away plan view of the passage.

Cut-away side view of the passage.

Scale

15m

While some sockets were found into which stones would have been inserted, there were not enough to provide conclusive evidence of a regularly spaced circle of stones. Having examined the cairn slip, some of which had been trapped under fallen orthostats, it was clear that the stone circle had been erected before the cairn had collapsed.

In 1962, while he was excavating the stone circle, O'Kelly found a series of pits underneath the standing stones. Carbon dating indicated that the pits, which are older than the stone circle, date from about a thousand years after the mound was constructed. More pits, part of the series discovered in 1962, were excavated in 1982 by David Sweetman, close to one of the standing stones. The holes are believed to have held timber uprights, forming a henge with a diameter of about 100 metres, the same size as the stone circle.

In 1967 O'Kelly made his most startling discovery. When the decorated stone above the entrance to the tomb, long regarded as a lintel, was excavated it was shown to be the roofstone of a box-like structure with a front opening, set back 2.5 metres from the entrance. It is 90 centimetres high, a metre wide, and measures 1.2 metres from front to back. O'Kelly had never encountered anything like it before (none of the other passage tombs in Brú na Bóinne have such a feature) and he called it a 'roof-box'. The back of the roofstone rested on the corbels in the passage in such a way as to create a narrow slit, which was closed with two small blocks of quartz that had clearly been moved out of the slit and back again on a regular basis. While he had been working on the site, various local people had told O'Kelly that the sun's rays entered the tomb at the winter solstice.

This discovery indicated that there could be some truth in this, so that winter,

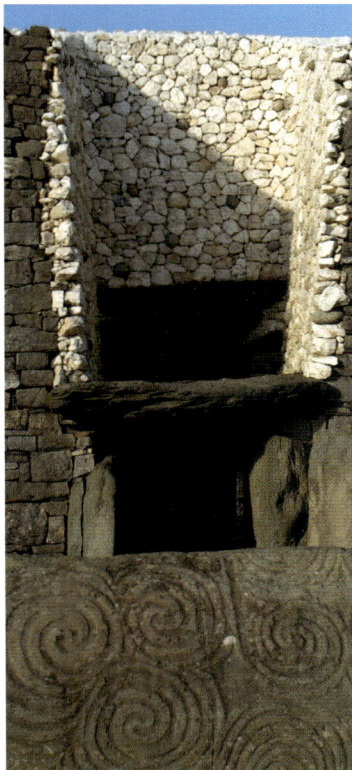

The entrance stone with the roof-box visible above.

he came to the site alone at the solstice, 21 December 1967, and entered the chamber at the back of the tomb just before the sun rose. Fortunately, the sky that morning was clear, and at precisely 8.58, the sun began to shine into the tomb. Michael O'Kelly became the first person for thousands of years to witness the spectacle that has made Newgrange internationally famous. He described the experience in vivid detail:

At exactly 8.54 hours GMT the top edge of the ball of the sun appeared above the local horizon and at 8.58 hours, the first pencil of direct sunlight shone through the roof-box and along the passage to reach across the tomb chamber floor as far as the front edge of the basin stone in the end recess. As the thin line of light widened to a 17 cm-band and swung across the chamber floor, the tomb was dramatically illuminated and various details of the side and end recesses could be clearly seen in the light reflected from the floor. At 9.09 hours, the 17-cm band of light began to

Newgrange roof-box. The entrance to Newgrange showing the entrance kerbstone K1 and roof-box. On the winter solstice the morning sun shines through.

narrow again and at exactly 9.15 hours, the direct beam was cut off from the tomb. For 17 minutes, therefore, at sunrise on the shortest day of the year, direct sunlight can enter Newgrange, not through the doorway, but through the specially contrived slit which lies under the roof-box at the outer end of the passage roof.

(O'Kelly, *Newgrange: Archaeology, Art and Legend, Thames & Hudson*, 1982)

The entrance to Newgrange on the morning of the winter solstice, which is marked by pagan celebrations.

The Finds at Newgrange

Over the centuries, various objects were found in the tomb; these were removed by those who found them and have ended up in private and public collections. A number of items of Roman jewellery found there are currently on display in the British Museum. By the time Michael O'Kelly was excavating the tomb in the 1960s and 1970s, there was little to find. He uncovered some stone marbles, four necklaces, two beads, some fragments of bone tools and some flint. The tomb revealed deposits of human bone, some burnt, some not, suggesting that both cremated remains and human corpses had been interred in the tomb. While there were also discoveries of animal remains, it seems likely that these would not have been placed there deliberately – animals would probably have found their way into the chamber and died there. Remains of pottery dating from the Beaker era were found outside the tomb – Michael O'Kelly believed that by 2000 BCE, Newgrange had fallen into disuse and people were 'squatting' in the area around it.

Subsequent Discoveries at Brú na Bóinne

In the very hot, dry summer of 2018, referred to as 'the summer of discoveries', the outline of other monuments began to appear in the flood plain below Newgrange, including what looks like a 100-metre diameter henge about 500 metres to the south-west of the Newgrange mound. A passage tomb about half the size of Newgrange was excavated at Dowth Hall by Clíodhna Ní Lionáin. During a dig directed by Matthew and Geraldine Stout, funded by the Royal Irish Academy, a linear monument was excavated near the mound, comprising pits and ditches that it is believed formed a cursus, possibly with a processional purpose. Some charcoal found at the bottom of one of the ditches was carbon dated to the Neolithic.

Drone photograph believed to be the footprint of a huge late Neolithic henge or enclosure. The markings are visible because of a prolonged drought at the time.

One of the most fascinating aspects of Brú na Bóinne is the art that was discovered there, both inside and outside the mounds. Out of a total of 900 decorated megalithic stones in Europe, 400 come from the tombs in the Brú na Bóinne complex, and more than a further 100 from another 37 tombs in the Boyne Valley. Some of the megalithic art in the valley is spectacular – that at Newgrange is considered to be the finest in Europe. Connections with the examples found here have been found at Neolithic sites in Brittany.

Spirals, lozenges (rectangles), chevrons (zigzags) and triangles are carved on the stones at

Newgrange in wonderful combinations. Claire O'Kelly, who carried out extensive research into the designs on the stones, found that 10 different decorative motifs were used (circles, spirals, arcs, serpentiforms, dots-in-circles, chevrons, lozenges, radials or star shapes, parallel lines and offsets or comb-devices), with the chevron and lozenge appearing most frequently. Most visitors to Newgrange will come away with the impression that there are many carvings of spirals, simply because they are very prominent, not least on the entrance stone. For many people these spirals represent the sun, although there is no evidence that this was the intention of the people who carved them. The image of the triple spiral has been incorporated in many modern jewellery and ceramic designs, and is often erroneously believed to be a Celtic symbol.

No one can say what the carvings mean, and there have been many theories, ranging from the idea that they are a simple depiction of the surrounding landscape (the spirals representing the Brú na Bóinne mounds, and the rectangles the fields created by the farmers), abstract representations of human faces, or astronomical images and cycles. It has even been suggested that they might be a kind of written

Decorated kerbstone at Knowth.

communication. Because it is thought that the Neolithic builders of the tombs may have used the constructions for shamanic rituals, there is a theory that some of the carvings were executed by people experiencing an altered state of consciousness, perhaps under the influence of hallucinogenic drugs. Whatever their purpose, many of the carvings are extremely beautiful, and, since their discovery, have frequently been replicated by artists and designers.

Given the tools the Neolithic people had at their disposal, the carvings are all the more remarkable. Today, incisions in stone are made using a tempered metal chisel and a hammer. The hardest materials available to the carvers at the Brú na Bóinne sites were flint and

porcellanite. The skill and determination of the creators of the carvings of at least three of the kerbstones is to be admired, and the difficulty of executing such accomplished works means that the carvings must have been extremely significant to the builders of the mounds.

Two different methods were used for carving the stones:

- The designs were incised, or etched, into the stone by dragging a sharply pointed implement across the surface. This produced shallow incisions.

- The carvings were picked or 'pocked' with a type of flint or quartz chisel to make much deeper incisions, such as those on the Newgrange entrance stone.

- Some deeper, 'pocked' incisions even produced a three-dimensional effect by creating the illusion of a relief, with the design appearing to stand out from the surrounding stone.

The Kerbstones

The stone that guards the entrance to the passage tomb at Newgrange (referred to by archaeologists as Kerbstone 1) is widely regarded as a Neolithic masterpiece without peer in Europe. It was the most important stone on the site – anyone entering the passage would have been obliged to climb over it (adjustments were made during the reconstruction process to make it easier for modern visitors to the tomb to gain access).

The stone is 3.5 metres in length, and the spirals, lozenges and diamonds that were incised on it were engraved while it was *in situ* at the entrance to the tomb (the carving stops at ground level and nothing has been carved on the underside

of the stone). There is a central groove that is mirrored by a groove on Kerbstone 52, which stands on the other side of the mound exactly opposite Kerbstone 1. The groove indicates the position of the rays of the solstice sunrise as they hit the roof-box. The dominant design on Kerbstone 1 is the spiral. There is a large triple spiral to the left of the central groove, and several spirals rotate from the groove on the right-hand side of the stone.

It is believed that Kerbstone 52 was carved, again *in situ*, by the artist who worked on the entrance stone, which led archaeologists to believe that, as at Knowth, there might be another passage behind it. Despite the best efforts of the archaeologists carrying out the investigations, one has not yet been discovered. There are the expected spirals and lozenges on the left-hand side of the central groove, but the designs on the

Newgrange entrance stone, K1, carved with the famous triple spiral design.

right are unusual, and would not seem out of place on an Inca or Aztec carving. The central groove is aligned with the summer solstitial sunset.

The only other stone that is fully decorated is Kerbstone 67, due north of the chamber, which can be seen if you walk around the mound in an anti-clockwise direction from the entrance. This stone is decorated mainly with chevrons and spirals.

The Roof-box

A carved, repeating pattern of lozenges was uncovered when the roof-box was first uncovered, giving rise to the belief that it was merely a lintel over the entrance to the tomb. William Wilde thought that *'its sculpture, both in design and execution, far exceeds any of the rude carvings … found within the cave; and as it could never have been intended to be concealed from view, it is most probable that*

Lozenges carved above the roof-box.

it decorates the entrance into some other chamber …'. Claire O'Kelly describes this carving as 'conscious ornamentation', probably carried out by the artists of Kerbstones 1, 52 and 57. It contrasts with what she describes as some of the not particularly well executed 'doodles' on some of the other stones.

Inside the Tomb

It is not only the stones at the exterior of the tomb that were decorated. Carvings are also to be found along the passage and in the chamber. The passage has 22 upright slabs, or orthostats, on the left-hand side and 21 on the right. The triple spiral is repeated on one of the stones in the passageway and on another in the chamber. Lozenges and zigzag patterns also appear on the orthostats and in the chamber.

Just inside the entrance to the tomb there are two horizontal picked stripes – the texture is such that the temptation to touch

Carvings on the Newgrange interior, with 19th C graffiti.

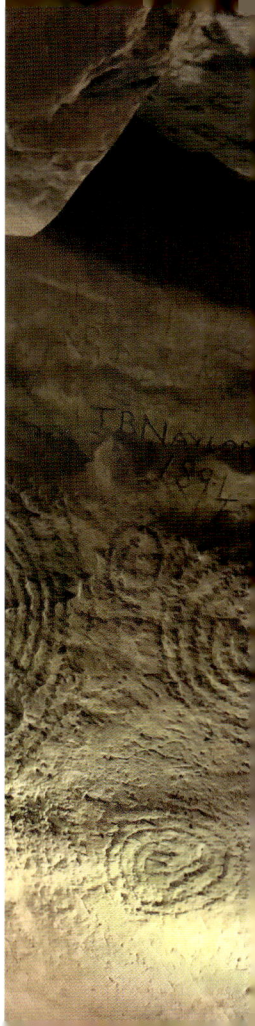

them, to run one's hands over them, is very strong. One of the most striking stones in the tomb is a little further up on the left-hand side of the passage, just as it bends into the chamber. The three spirals with a central lozenge seem to represent a face, and the zigzag patterns surrounding it look a bit like a wild hairstyle.

Described by antiquarian George Coffey as one of the most famous stones at Newgrange, this has come to be known as the 'Guardian Stone', and it has been speculated that it represents a spirit who watches over the remains of the dead placed in the tomb. Others think it might have been placed there to frighten unwelcome visitors to the tomb.

The last stone on the right became the stuff of legend. It has several horizontal picked bands, and indentations and lozenges were carved above these. A long-ago caretaker of the tomb told visiting children that the bands were the imprint of a giant's rib cage, formed as he squeezed through the passage into the chamber. The shapes created by the shadows on this stone draw the visitor forward into the chamber.

Once inside the chamber, the most extravagant decoration appears on the roofstone of the recess on the

Triple spiral on a wall.

right, swirling across the entire surface. At the back of the chamber, on a stone that had fallen forward slightly, there is a carving of a triple spiral. Local lore suggested that the sunlight fell on this stone on several days of the year, but little credibility was attached to this tale until Michael O'Kelly discovered the roof-box and found that the dawn rays of the sun at the winter solstice did indeed enter the chamber and would have fallen on the stone (changes in the earth's orbit mean that they no longer reach that far). In the right-hand recess, there are two stone basins, one of which is beautifully decorated.

The Hidden Art at Newgrange

As the excavation proceeded it became apparent that not all of the decoration on the stones was visible. Some of the stones had been carved on their undersides, or on the sides that faced inwards. It has been suggested that the act of carving the stone was of importance, but that what was carved was not meant to be seen; perhaps it was meant to be visible only to the dead whose remains were found in the tomb, or to the gods. There is another theory that the

stones were carved elsewhere for some other monument, and were brought to Newgrange where they were repurposed.

Michael O'Kelly had a more mundane explanation – the art was never intended to be hidden, but when the decorated stones were put in position the fit was sometimes better if they were turned to face inwards.

Although we will never know what the carvings at Newgrange mean, it is impossible not to be impressed by their beauty and the skill of the people who created them.

Stones with carved decorations.

Restoring Newgrange to its former glory required ingenuity, a high level of engineering and architectural knowledge and skill, and a great deal of hard work. There has been much criticism of the reconstruction of the mound, but Michael O'Kelly knew that visitors would continue to visit the site in their tens of thousands annually, so any reconstruction had to withstand their intrusion. The degradation of the site over the millennia showed the extent to which the stones were susceptible to being damaged by human traffic.

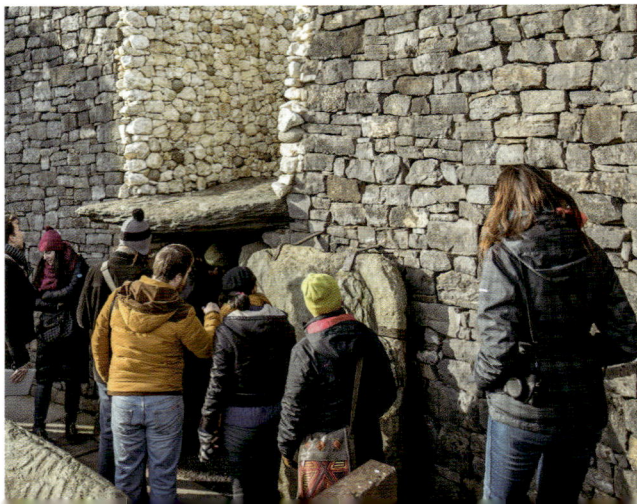

The Kerbstones

The restoration began with the kerbstones, many of which had fallen over. These were straightened and secured, and in cases where the carvings were facing inwards, the decoration was photographed before the stones were adjusted. Two kerbstones split by tree roots were glued together.

The Revetment Wall and the Entrance

A civil engineer who was asked to calculate the dimensions of the quartz revetment wall so that it could be reconstructed, concluded that it must have been in the region of three metres high. Rebuilding it as a dry stone wall would result in its eventual collapse, so it was decided to reinforce the edge of the mound to prevent future instability. A reinforced concrete retaining wall was built behind the

Construction of the retaining wall.

kerbstones, and the quartz and granite stones were fixed to this using a mortar that is invisible to the onlooker.

As it wasn't clear what the entrance had looked like (although O'Kelly assumed that as none of the cairn material had been scattered in the entrance, the revetment wall must have turned inwards towards the entrance and over the sides and lintel of the roof-box), and because of the need to accommodate visitors, the revetment wall was brought as far as the outer ends of the two kerbstones adjoining the entrance stone on either side. It is connected on each side of the entrance to a curved grey limestone wall that extends as far as the entrance. The wall immediately above the roof-box is faced with white quartz.

Wooden stairs were built over the two kerbstones that adjoin the entrance stone, to provide access to visitors in a way that avoids further damage to the stones.

Work on the site.

Drainage

To preclude the possibility that rainwater would pool behind the revetment wall, drains were built in the north and south sides of the mound, the two drainpipes running from just above the level of the turve layer and out under the foundation of the wall between each pair of kerbstones into a series of sinkholes. In order to make sure that the system worked, the operation of the drains was observed for a winter before the cairn was rebuilt. A sheet of heavy-duty plastic was placed under the cairn stones when they were being replaced, to prevent organic material washing down and blocking the drains. A thin layer of cement now separates the different layers of the cairn materials. A drain was also installed in the passage to take away water that welled up from a natural spring under one of the orthostats. Unfortunately, in the 1990s some of the drainage holes in the retaining wall silted up and pressure built up at the back of the mound, so remedial works had to be carried out.

The Passage

Because many of the orthostats in the passage were leaning inwards, they were damaged by being repeatedly rubbed against by visitors making their way through to the chamber. In order to straighten them, the roof-box and the roof stones at the front of the passage had to be removed – each stone was numbered before removal so that it could be replaced correctly. One of the roof slabs was shattered, so it was replaced with a limestone slab, as were two corbels where the decoration was on the reverse of the stone – these were removed and brought to the National Museum in Dublin, where they are now on display. The passage and the roof were enclosed in a concrete tunnel to enable them to withstand the pressure of the restored cairn. The tunnel is high enough to allow access through an opening to inspect the roof slabs. The roof-box was reinstated, in a somewhat higher position than before. Electricity was installed in the passage to provide light.

The passage had a closing stone, which had ended up lying flat on the ground outside the entrance, its surface

The passage at Newgrange.

worn smooth by generations of feet. This was placed upright to the east of the entrance, the worn side turned inwards.

The Chamber

Some of the corbelled roof stones in the chamber had cracked and the roof was no longer waterproof. Water dripped into the stone basin in the east recess, something that had been commented on by that very early investigator of the mound, Edward Lhwyd. A pyramid of concrete pipes was built over the chamber, covered in a concrete roof of thin overlapping slabs, and this solved the problem of dripping water.

Aspects of Michael O'Kelly's reconstruction have been heavily criticised, including the revetment wall and the roof-box. However, the roof-box works, in that it permits the rays of the winter solstice sun to enter the chamber. As for the wall, without mortar it was bound to collapse – perhaps the builders of the mound had simply failed to anticipate this. Whatever the truth of the matter, the reconstructed monument, with its shining white quartz revetment wall, now dominates the surrounding landscape. It is impossible not to be awed.

Newgrange is the most visited monument in Ireland, with upwards of 200,000 visitors annually. The only access to Newgrange and Knowth is from the purpose-built Brú na Bóinne vistor complex at Donore, County Meath. There are tours available from Dublin, but it's easy to make the trip independently.

Once you pass through the wonderfully crafted gates – they replicate some of the designs on the entrance stone at the Newgrange mound – at the entrance to the complex on the south side of the River Boyne, you enter a different world. Built low so as not to intrude on the landscape, its curvilinear outline suggesting the shapes of the Brú na Bóinne monuments, the grass-roofed visitor centre opened in 1997. It is the gateway to a fascinating journey into the Neolithic world.

There is ample parking, and signs direct you along the short walk to the visitor centre, which is wheelchair accessible. The centre has all the usual amenities – tea room, toilets, picnic facilities, shop and tourist information.

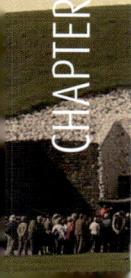

The exhibition at the centre provides an immersive experience of the sights and sounds of the Neolithic world of the Brú na Bóinne. There are visual and written descriptions of the people who built the tombs, the tools they used, and the kind of lives they led. Where they lived, how they dressed, the food they ate and the domestic utensils and weaponry they used are all detailed.

Much of the exhibition is dedicated to the construction and art of the monument. An audio-visual display explains the different solar alignments of the Boyne Valley monuments and shows what happens in the chamber at Newgrange during the solstice sunrise. There is a full-sized walk-through replica of part of the Newgrange passage.

The exhibition whets the appetite for a guided tour of the real thing. A short walk across a bridge over the River Boyne brings you to the bus stop, from where a shuttle bus will take you, first to Newgrange, and then on to Knowth, with ample time allowed to view the monuments.

As you climb up the hill to Newgrange and see, for the first time, the size of the kerbstones that surround the monument, you will be awed by the determination and ingenuity of the people who brought them there. Before

Knowth

the restoration of the monument anyone entering the passage had to climb over the entrance stone, but limestone wings installed in the 1970s allowed for the erection of steps, which make it easier for the visitor and prevent further damage to the stones. A guide leads you into the chamber at Newgrange for a light display that simulates the solstice event.

The mound at Knowth was constructed on a hill that is slightly higher than that at Newgrange. There are steps to the top of the mound – the glorious panoramic view of the countryside (which would have been covered in woodland in the Neolithic era) makes you realise why our ancestors would have built such an important burial place there.

You'll need at least 45 minutes in the visitor centre to get the most out of the exhibition, and the tour of Newgrange and Knowth takes about 3 hours altogether. The visitor centre is open all year round, except from 24–27 December each year. Visitor numbers are limited, so it's a good idea to book in advance.

The eastern passage of Knowth is 40 metres in length, making it the longest megalithic passage in Western Europe.

Many people dream of witnessing the solstice event, but this is limited to a very few people each year. A lottery selects 50 people (there are around 30,000 entries each year), each of whom can take one other person with them on the appointed day. Anyone else is welcome to stand outside the tomb at the appropriate time. Applications can be made by email to: BrunaBoinne@opw.ie
Make sure to supply your name, address, country of residence, telephone number and email address.

The sunrise enters the tomb on six days – 18–23 December – at 8.58 am, depending on the weather. Many a lucky lottery winner has stood in a dark chamber on a cloudy day!

At sunrise during the winter solstice a beam of light illuminates the central chamber at Newgrange.

CHAPTER 9 The Boyne Valley

The Boyne Valley Scenic Drive

One of the best ways to take in all that the Boyne Valley has to offer is to follow the route of the Boyne Valley Scenic Drive, which will take you through beautiful landscapes steeped in history and mythology. You will travel a route along 200 kilometres of wonderful scenery in the beautiful counties of Meath – the Royal County – and Louth, and visit ancient monuments, Christian monastic settlements, stately homes and gardens, and battle sites. This is Ireland's Ancient East, an area that holds the key to almost 9,000 years of history, from the Neolithic era until modern times. The route is easy to follow at your own pace – there are brown directional signs with the image of a spiral indicating the route to each of the monuments and attractions; or download the free App for iPhone/Android at activeme.ie.

Loughcrew

IRELAND

Donaghmore
Round Tower

Kells

Northern
Ireland

Ulster

Ulster

Connacht

Leinster

ATHBOY

NAVAN

Munster

LEINSTER

Trim Castle

River Boyne

Louth

Long-
ford

Meath

Westmeath

Offaly

Kildare

Dublin

Laois

Wicklow

TRIM

Carlow

Hill of Tara

Kilkenny

Wexford

N

NW   NE

W   E

SW   SE

S

DUBLIN

Mellifont Abbey

Monasterboice

Newgrange

SLANE

DROGHEDA

Battle of the Boyne

DULEEK

JULIANSTOWN

Duleek Abbey

DUBLIN

DUBLIN

A little to the north of Newgrange and about eight kilometres west of Drogheda (see page 134) is the larger tomb complex of Knowth. It is believed to have been constructed in 3200 BCE and has one large mound, surrounded by 17 smaller satellite mounds, some of which are older than the main mound. There is a reconstruction of a Bronze Age wooden henge near the entrance to the east passage. The large mound is 12 metres high and 67 metres in diameter. Steps have been installed to make it easy to access the top of the mound, from which there are panoramic views across several counties. On a clear day you can see the Wicklow mountains.

Knowth Passage Tomb

The Celtic name for Knowth is Cnobga, thought to derive from Cnoc Bua or Cnoc Buí (Hill of Bua or Buí). Buí was the so-called 'hag' of Beara, the earth goddess. She is also associated with death, so would have had a strong connection with a burial mound in the Celtic imagination.

Unlike Newgrange, the main mound at Knowth has two separate passages, accessed from opposite sides of the tomb and almost meeting in the centre of the mound. Each passage leads to a burial chamber. The larger eastern passage was the second one to be discovered by Professor George Eogan, who led the excavation of the site in the early 1960s. At 40 metres, the passage is twice the length of that at Newgrange, and, like the Newgrange passage, it ends in a cruciform chamber with three recesses, each containing stone basins in which the cremated remains of the dead were placed. The right-hand recess is larger and highly decorated with stone carvings. The western passage ends in a single chamber. The eastern chamber, as at Newgrange, has a corbelled roof, while the roof of the western chamber is a single flat stone, two metres in length.

The base of the exterior of the main mound is enclosed by 127 retaining kerbstones, almost all of which are engraved, some elaborately carved. Three are missing, and concrete placeholders have been inserted to fill the gaps.

It was believed that the passages and chambers at Knowth were originally aligned with the spring and autumn equinoxes, but that alignment is no longer in evidence. The original passages were destroyed or distorted, and changes in the earth's movements over the millennia may have had an impact on when the sun's rays hit any particular spot. It is now thought that Knowth was aligned with the moon rather than the sun.

Kerbstone with carved spirals.

At some point before the Bronze Age the mound slipped, closing off the entrances to both passages. Remains of graves have been found at the site – those interred in them were placed in small stone coffins called cists.

Later, during the early Christian period (around 450 AD) Knowth was the royal seat of the kings of Brega, the capital of the kingdom of Northern Brega, which included the Hill of Tara (see page 234). At this time a hill fort was built on top of the mound and two defensive ditches were dug, one at the top of the mound and the other behind the kerbstones at the base. There are Christian graffiti and ogham markings on the stones in the eastern chamber.

Aerial view of Tara.

A Norman soldier.

From the 8th to the 12th centuries the mound was the central point of quite a large and well-defended habitation. Artifacts from this period have been discovered, along with the remains of houses and a souterrain network of tunnels, used for storage and refuge.

The Normans used the mound at Knowth as a motte and bailey in the 12th century, taking advantage of its dominant position in the landscape, and the monks of the Cistercian foundation at Mellifont Abbey (see page 176) used it as a farm, with stone walls enclosing stone buildings. These fell into disuse when Henry VIII dissolved the monasteries, and the site was turned over to agriculture.

Knowth was acquired by the state in 1939. Excavations were carried out in 1941 and again in 1962, the latter coinciding with the start date of Michael O'Kelly's excavation of Newgrange. In 1967 and 1968 the entrances to the two passages were discovered.

The Art at Knowth

The decorated stones in Knowth form more than a quarter of the repository of megalithic art in Western Europe and almost half of the total of that found in Irish passage tombs. As at Newgrange, many of the decorated faces of the stones were turned inwards, and this gave rise to the idea of hidden art. It is unclear whether there was some intention to hide the art, or whether the stones were recycled from other sites. George Eogan suggested that the mound at Knowth may have been constructed on top of an earlier site, and some of the carved stones there might have been recycled from that site. He also believed that the stones at Newgrange might have been similarly recycled from another, earlier site.

Spiral decorations at Knowth.

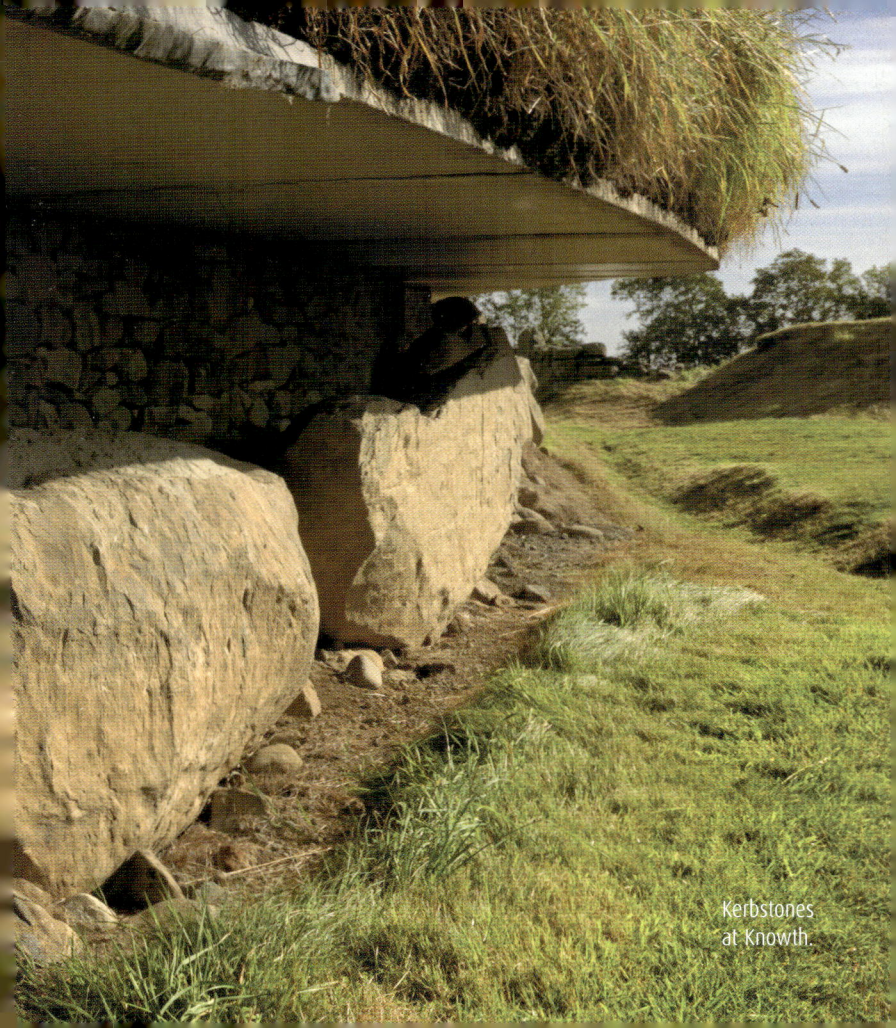

Kerbstones at Knowth.

The theory that Knowth had a lunar alignment is reinforced by the type of decoration that was found on the stones there, which is very different from the carvings at Newgrange. Here, the familiar spirals and lozenges that are prevalent in Neolithic art are supplemented with motifs that include curved lines and circles or discs, which suggest that they represent the phases and cycles of the moon – the progression of the carvings from kerbstone to kerbstone gives the idea that they depict the waxing and waning of the moon. Along the eastern passage there is a stone that seems to have been created with the same intention as the Guardian Stone at Newgrange (see page 90). Eogan referred to it as the 'ghostly guardian' of the burial chamber – it also looks very like an owl. A dressed stone on the left-hand side of the chamber has retained a smooth animal-shaped portion at its centre. A beautifully decorated stone basin (also known as the Dagda's cauldron), with a diameter of more than a metre, was discovered in one of the recesses of the chamber. It is still there, as it was too large to be moved out of the chamber.

The most significant artifact discovered at any of the Boyne Valley Neolithic sites is the Knowth Macehead, an exquisitely carved ceremonial flint macehead, regarded as one of the finest Neolithic artifacts in Western Europe, the carving of which would have required extraordinary precision and skill. It seems to represent a human head, with spirals suggesting eyes, and a mouth-like hole into which the mace's wooden handle would have slotted. It was carved out of a single piece of grey flint streaked with brown, and at just under 8 cm in length it is small enough to fit in the palm of the hand. The macehead is now housed in the National Museum of Ireland, Kildare Street, Dublin.

Knowth is open to the public but access is possible only from the Brú na Bóinne Visitor Centre. It isn't possible to enter the mound, as at Newgrange, but you can go into a small chamber at the beginning of one of the passages, which affords a good view down the passageway.

The Knowth Macehead.

Dowth Passage Tomb

Although Dowth is as large as neighbouring Newgrange and Knowth, it is the least well known of the three principal Neolithic passage tombs in the Boyne Valley. Unlike Newgrange, Dowth has two chambers (Dowth North and Dowth South), both on the west side of the mound, set 20 metres apart, each with a low corbelled roof. The mound hasn't been professionally excavated, although various amateurs carried out damaging investigations over the years, particularly in the mid-19th century. The two chambers are lower than those at Newgrange and Knowth, so they would have been more difficult to access. Dowth North is the more complex of the two, with a longer passage, at 14 metres, a cruciform chamber, as at Newgrange and Knowth, and two small chambers leading off the right-hand recess. The passage in Dowth South is only 3.5 metres long, but although both passages are shorter than the Newgrange passage, the chambers are the same size.

View of Dowth. The mound is believed to have been more conical in shape before its ceiling collapsed in 1847 after amateur excavations.

Archaeologists believe that Dowth was constructed more recently than Newgrange, with estimates ranging from 3200–2000 BCE. It may have been ransacked by Danes looking for treasure. Dowth North collapsed completely during the construction of a medieval souterrain, built to provide storage space and as a place of refuge during times of invasion. Amateur excavations in 1847 caused the centre of the 50-metre high mound to cave in, forming a huge crater-like depression. Before that, it was, according to William Wilde, 'more conical' than Newgrange, though 'not so broad at the base'.

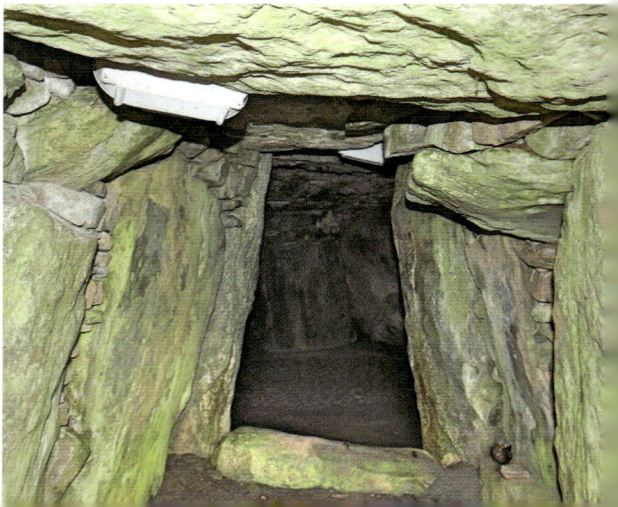

Inside the passage leading to the central chamber of Dowth.

Wilde was one of the lucky few to gain access to the interior of the mound. In 1848 he wrote that:

Of the internal arrangement of this huge cairn little, until very recently, was known beyond the fact that … there appeared here the remains of two passages in a very ruinous state and completely stopped up, neither of which however, seemed to have conducted towards a grand central chamber … a sepulchral chamber of a quadrangular form, the stones of which bear a great variety of carving … has been discovered upon the southern side of the mound.

(William Wilde, *The Boyne and the Blackwater*, 1848)

Sir William Wilde, 1815–1876, was one of the only few who had gained access to the interior of Dowth Passage Tomb.

Spiral decoration at Dowth.

Like Newgrange and Knowth, Dowth is aligned with the sun at the winter solstice, although in the case of Dowth, it is the sunset rather than the sunrise that permeates the interior of Dowth South. From the beginning of October until the end of February each year the rays of the sun find their way into the chamber. By sunset on the winter solstice, they shine directly into the back of the chamber.

The entrance to Dowth North was destroyed several centuries ago when a souterrain was built right across it, so it is impossible to ascertain when, or even whether, the sun's rays hit the back of the chamber.

The Art at Dowth

The patterns carved into the stones at Dowth are fewer, simpler and seem more random than those executed at both Newgrange and Knowth. They include spirals, chevrons and lozenges, and there are several instances of what seems to be a depiction of the sun's rays. One of these, what appears to be a sunburst, has been carved on the right-hand side of the passageway at Dowth North, an indication that the tomb, like Dowth South, was designed to welcome the sun.

There is no public access to the interior of the mound. However, while access to the other two Neolithic sites is only through the Brú na Bóinne visitor centre, Dowth is not on the official itinerary and visitors can drive there directly. As it hasn't been developed as a tourist site, it's usually very quiet and there are wonderful views of the surrounding countryside from the top of the mound. William Wilde records that John, sixth Viscount Netterville of Dowth, renowned for his eccentric ways, built a tea-house (no longer in existence) on top of the mound to take advantage of the sweeping vista. However, archaeologist and art historian Peter Harbison says that the structure built by Netterville, a Protestant, was a 'temple' from which he could follow the Catholic Mass at the nearby chapel each Sunday. Whatever its purpose, it was destroyed by the mid-19th century excavators and nothing of the structure remains.

Duleek Abbey

In 450 St Patrick established an episcopal seat at Duleek, paving the way for the foundation of a monastery there by St Cianan. Cianan built a church, which dates from before 489, said to be the first stone church built in Ireland. The name Duleek derives from 'Daimhliag Cianan', meaning Cianan's stone church. The ruins of the church can be seen opposite the ruins of the abbey, St Mary's, that was built on the site in the 12th century when the church and its lands were granted to the Augustinians by Hugh de Lacy, Lord of Meath. Inside the ruin of the abbey church there are some fascinating tombs, including a beautifully decorated box tomb, dating from the 17th century, and the effigial tomb slab of a 17th-century bishop of Meath, James Cusack. There are two high crosses, one of which, dating from the ninth century, is carved with scenes from the life of the Virgin Mary and the symbols of the four evangelists.

During the 9th and 10th centuries Duleek was attacked by Vikings. The small community of monks survived, and it seems fitting that in 1014, after the Battle of Clontarf in which the Vikings were decisively beaten, Brian Ború's body lay in state at

Duleek before being brought for burial to Armagh.

Don't miss the 15th-century bell tower, the north wall of which contains a 'ghostly' impression of an earlier round tower. An opening would have led from the bell tower to the round tower, so the round tower must have been demolished at some point after the construction of the later bell tower.

St Marys' Abbey, Duleek, one of Ireland's oldest churches.

St Mary's Abbey,
Duleek.

Drogheda

Drogheda (Droichead Átha, 'the bridge at the ford'), the gateway to the Boyne Valley, was one of the most important and best-fortified towns in Ireland during the medieval period, when it was included in the area of the Pale. Situated a few miles inland from the actual mouth of the river Boyne at Mornington, Drogheda had one of the most strategic locations in the country, and the town controlled the political and commercial life of the area. It hosted several meetings of the Irish Parliament, which at that time moved around the country from location to location. Drogheda is still one of Ireland's larger towns.

Hugh de Lacy Bridge was named in honour of the founder of Drogheda. Also known as The Millennium Bridge.

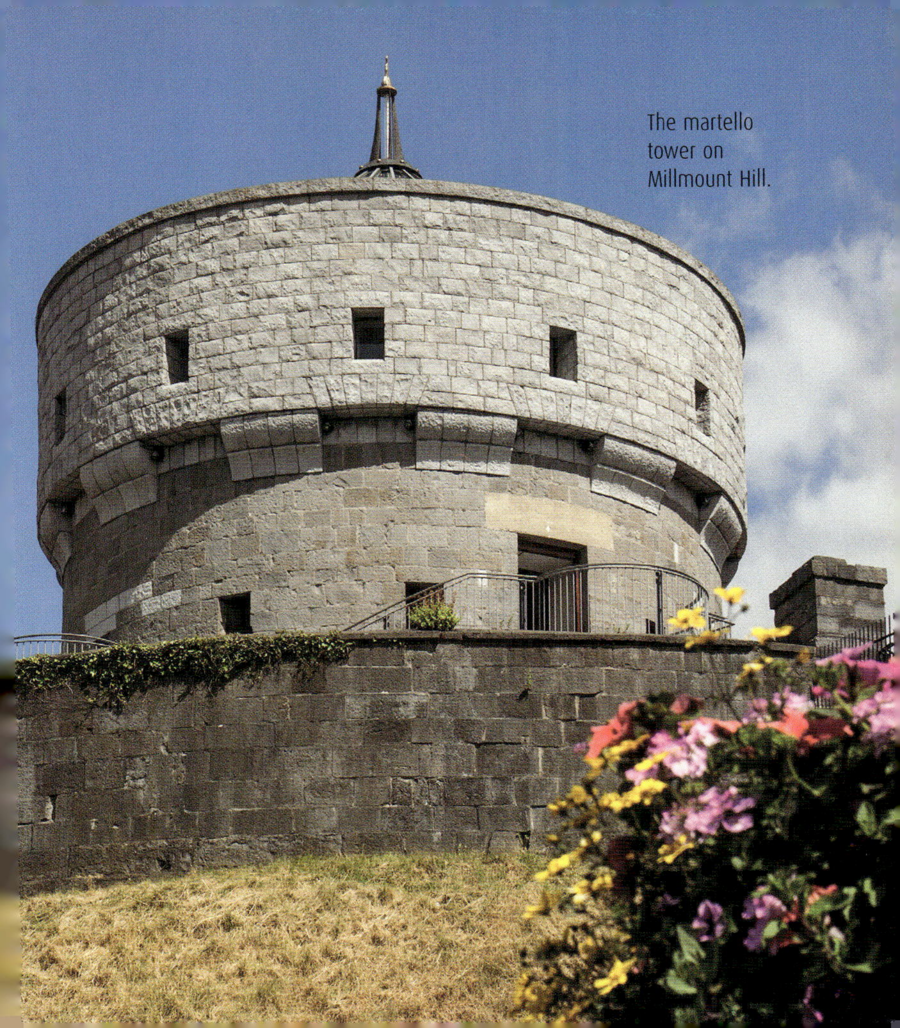

The martello tower on Millmount Hill.

Magdalene Tower, a landmark located at the highest point of the northern part of Drogheda.

One of the earliest constructions in Drogheda was the Anglo-Norman Hugh de Lacy's motte and bailey at Millmount (see page 164). Drogheda developed around the fort and in 1194 de Lacy's son Walter granted it the first town charter in Ireland. At the time of its incorporation it straddled two modern counties – Meath (the Lordship of Meath was much larger than the modern county) and Louth (then known as Oriel).

Drogheda's 17th-century history was the most turbulent of its existence; the town endured two sieges in less than a decade. Its substantial grain store alone would have been sufficient reason for an army on the move to attempt to breach its walls. In 1641 a rebellion began in Ulster, when Catholic landowners rose up against the Protestants of the Ulster Plantation. It was the beginning of the Confederate Wars. Atrocities were committed on both sides and the conflict soon spread to other parts of Ireland. Having won a victory over a government army at Julianstown, County Meath, on 29 November 1641, the rebels, led by Phelim O'Neill, laid siege to Drogheda. The town garrison held them off, and when some rebels eventually found a way into the town they were taken prisoner. The garrison sent for reinforcements while the rebels made one last unsuccessful attempt to take the town. The reinforcements arrived and crushed the rebels, driving them back into Ulster.

Sir Phelim O'Neill led the siege on Drogheda in 1641.

When England was declared a protectorate under Oliver Cromwell in 1649, royalists in Ireland were quick to rebel. In August 1649, Cromwell landed in Dublin, along with an army of 15,000 troops, to quell the rebellion. The royalists retreated to fortified towns, including Wexford and Drogheda. The royalist governor of Drogheda, Sir Arthur Aston, had complete confidence in the strength of the town's defences – he famously said that anyone who could take Drogheda could take

Oliver Cromwell, leading the storming party during the siege of Drogheda, 1649.

hell – and when Cromwell arrived on 3 September with an order to surrender, Aston refused to entertain it. On 10 September, Cromwell launched an artillery attack, but failed to breach the walls to the extent that his troops could gain entry to the town. However, although the walls were high, they had been built before the cannon arrived in Europe, and they were too thin to withstand the artillery assault for long. The following day, Cromwell breached the walls and 6,000 of his troops poured into the town. They carried out one of the worst massacres ever to take place in Ireland. Catholic priests and monks and any people who were deemed combatants were slaughtered in the streets, and a group that had barricaded themselves in St Peter's Church were burnt to death when the church was set on fire by the invading troops. A cavalry regiment lay in wait outside the town, mopping up those who managed to escape. The death toll was 2,000. Having taken Drogheda, Cromwell continued his bloody campaign in the rest of Ireland, earning him an everlasting reputation as a monster, especially in those towns that were subjected to the worst horrors of his campaign.

The St Laurence Gate

This twin-towered barbican in the east wall of the town shows the scale of the massive walls that surrounded the town, and is considered to be the most impressive survivor of defensive town walls in the British Isles. The top of the barbican was a superb vantage point from which to surveil the boats coming up the river; the view from there extends for several miles. The gate is an iconic image for the people of Drogheda – it's a prominent feature of the town's coat of arms. In 2017 the townspeople voted to close the area around it to traffic, as a protective measure.

St Peter's Church

The spire of this French Gothic-style church dominates the skyline of the town. Built in 1884, it replaced an earlier church built in 1793 by renowned architect Francis Johnson. Its central position is unusual in an Irish town – under the Penal Laws, Catholic churches had to be built outside the city walls or at least in a side street. In 1681, Drogheda's most famous son, Oliver Plunkett, the Catholic archbishop of Armagh, was executed as a traitor at Tyburn in London, having been convicted on foot of a trumped-up charge of involvement in a 'Popish plot' at a time when anti-Catholicism was rife. He was beatified in 1920 and his relics were brought to Drogheda. His head is now on display in a glass-fronted reliquary in St Peter's. When he was canonised in 1975, St Oliver became the first new Irish saint for almost 700 years.

St Peter's Catholic Church.

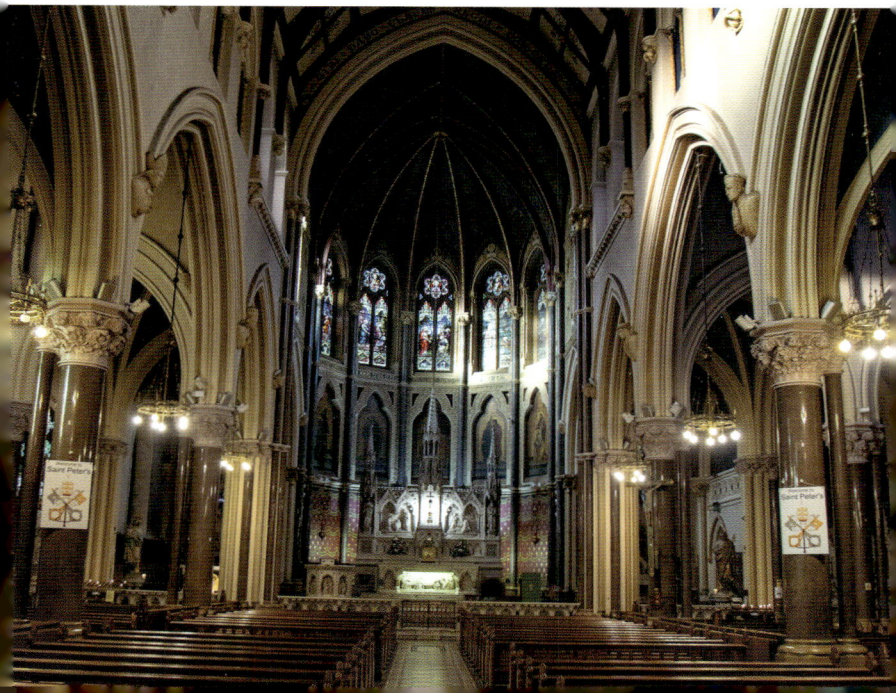

The Magdalene Tower

This tower is a Drogheda landmark, rising high above the surrounding landscape. It dates from the 14th century, and was constructed as the belfry in a large Dominican friary that was built there in the 13th century. The friary was the location for the submission of the great northern lords to King Richard II in 1395.

Famine Memorial

The population of Drogheda was devastated by the Great Famine of the mid-19th century and a small memorial on West Street acknowledges the contribution of £1,000 and several shiploads of grain to the starving people of Ireland in 1847. The Sultan of the Ottoman Empire sent several shiploads of grain

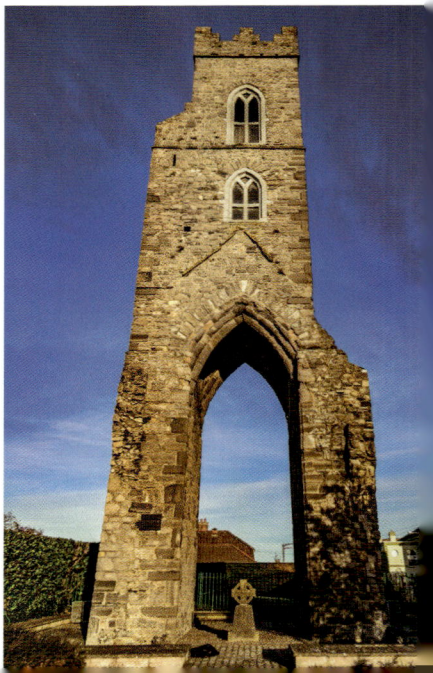

The Magdalene Tower, a landmark located at the highest point of the northern part of Drogheda.

to Dublin, where they were refused permission to dock by an embarrassed administration. The ships proceeded up the coast to Drogheda, where the grain was unloaded and distributed. In 1995 a plaque on the Westcourt Hotel was unveiled:

The Great Irish Famine of 1847 —
In remembrance and recognition
of the generosity of the People of Turkey
towards the People of Ireland.

The Drogheda coat of arms incorporates a star and crescent, which is believed to be a tribute to the Sultan's generosity in Ireland's time of great need.

The Sultan of the Ottoman Empire, Abdülmecid I, 1823–1861, sent shiploads of grain to be unloaded at Drogheda, after being refused docking at Dublin.

The Battle of the Boyne

Battle of the Boyne, Benjamin West (1738-1820).

The European Context

The Battle of the Boyne, fought on 1 July 1690 on the banks of the river just to the west of Drogheda, is usually described as the most important battle in Irish history. But it was also, almost uniquely for an Irish battle, one that had European consequences as well as local ones.

Ireland was a distant but significant theatre in the greater European conflict known as the War of the League of Augsburg. The contending forces were the French under Louis XIV, then at the apogee of his power and looking to establish absolute hegemony in Europe, and a coalition of interests put together by William of Orange, the *stadhouder* of the Calvinist Netherlands and recently installed as King William III of England. France was a mortal threat to Dutch independence and by some distance the most formidable military power in Europe. Its aggressive ambitions could be stopped only by a coalition: no other power could beat France on its own.

It was the same logic that had defeated Spain, when she was in the ascendant, and that would defeat France again in the Revolutionary and Napoleonic wars; it took the sixth of six coalitions to see off Napoleon. Later, it took different coalitions to defeat Germany, by then the greatest European power, in the two world wars.

The anti-French alliance – the League of Augsburg – was formed in late 1689 between the Netherlands, the Austrians and the English, with one or two others joining later. The immediate pretext was a French invasion of the Rhineland the previous year. No one felt safe from French aggression and the coalition of significant military powers made every sense: indeed, it offered the only possible route

King Louis XIV of France.

to victory, or at least to a draw. And a draw is more or less what it achieved, as the French realised that while they could not be defeated, they could not win outright either. In all, this war consumed nine long years before a treaty was signed which brought it to an end.

In the meantime, Ireland had entered the European picture. The legitimate king of England from 1685 had been James II, a Stuart. But James was a Catholic, and England was very Protestant. Only five years before he ascended the throne, the country had been thrown in civil turmoil by the so-called Popish Plot, an outbreak of anti-Catholic hysteria that, among other things, brought the judicial murder of the blameless Oliver Plunkett. So the Catholic James was an object of suspicion in Protestant England, and suspicion turned to outright alarm when the queen produced a legitimate male heir, thus securing the Catholic succession.

A *coup d'état* in the Protestant interest followed in which James was usurped and his place taken by his son-in-law, the Dutch *stadhouder*, who now became King William III of England and Ireland. This act was subsequently dignified by the name of the Glorious Revolution.

Portrait of King James II, who was usurped by William III. Peter Lely, c. 1655-70.

Richard Talbot, (c.1625–1691), first earl of Tyrconnell. Watercolour portrait by John Bulfinch (d.1728) after a painting by Kneller.

The Irish Context

After his deposition James fled to France, there to seek the protection of Louis XIV. His cause in England was hopelessly lost, but he had a natural constituency in Ireland. In most of the island outside Ulster, the Reformation had failed and Catholicism was still the professed religion of elite and commoner alike. Moreover, one of the leading men of the Catholic elite, Richard Talbot, had been elevated to the rank of duke of Tyrconnell by James II and given funds to strengthen his army. He replaced Protestants with Catholics in the Irish administration. The English watched this with apprehension, fearing that Tyrconnell's Catholic army could be deployed in England in support of a Stuart restoration. In Ireland, Tyrconnell held out the promise of restoring Catholics to the lands that they had lost 30 years earlier under Cromwell.

Tyrconnell's troops now moved towards Ulster. The entire province seemed at their mercy, until the heroic defence of the besieged town of Derry – 105 days of successful defiance which was seared into the collective memory of Ulster Protestants – slowed the Jacobite advance, buying time and bringing some important visitors to this remote theatre of the War of the League of Augsburg.

Siege of Derry, 1689.

First to arrive was King James himself, despatched to Ireland to provide a distraction while Louis concentrated on defeating the Dutch. In the process, it was hoped that James could recover his Irish kingdom, even if his English one was gone. A Jacobite Ireland would be in Louis's pocket, a handy bonus at England's back door.

James was very well received in Ireland, and every honour was shown him. French troops were sent to augment his army and soon James had control of the entire island – except for the hold-out at Derry. The little town had no right to hold out, and conditions there were frightful, but hold out it did. The Jacobites did not have enough siege guns but they managed to throw a boom across the River Foyle to stop relief supplies reaching Derry. Williamite ships eventually broke the boom and Derry was relieved on 31 July 1689.

The second visitor was a veteran French Huguenot general now in the service of King William, the duke of Schomberg. He landed at Carrickfergus, on the far side of Ulster, in June 1689 and soon he was joined by the third and most important visitor, William himself. He assembled an army in London, marched it up to Chester, and sailed for Carrickfergus, where he landed on 14 June 1690. He met up with Schomberg and other military captains in Belfast.

The two kings, James and William, were now in Ireland, one the client of France and the other its mortal enemy. It was only a matter of time before they met in battle – which brings us to the Boyne.

The armed merchant ship *Mountjoy* breaks through the defensive boom to relieve the Siege of Derry.

The Battle of the Boyne

King James's troops, known as Jacobites after the Latin for James, Jacobus, took possession of the south bank of the river at Oldbridge. In due time, the Williamite army arrived and took up its position on the north bank. It was an international force, as befitted its coalition status. It was bigger than the Jacobite army opposite – 36,000 men as against 25,000 – and many of the troops were battle-hardened from previous European conflicts. Moreover, it had a clear advantage in gunnery and artillery.

On the eve of the battle, King William reconnoitred his position, but this put him in plain view of the Jacobite troops on the south bank, one of whom took a pot shot at William and drew blood. It was only a flesh wound but it was rather too close for comfort.

The battle was fought the following day, 1 July 1690 (old style, adjusted to 12 July with the later introduction of the Gregorian calendar). William's battle plan worked as well as he could have hoped. He despatched some of his troops to the west, upstream, to assume position at the ford at Slane. The Boyne is tidal up to Slane and the ford there would have represented a plausible battle site. This manoeuvre convinced James that it was there that William proposed to give battle. Accordingly, he sent his best troops, the French, to defend the ford.

This suited William very well, for James had fallen into a trap. Once the French were gone and the tide had begun to ebb, the Williamites opened up a frontal artillery barrage across the river at Oldbridge before sending their most trusted Dutch troops directly across the river. Even on the ebb tide, the river was still high and they had to wade across with their arms held above their heads, an obviously dangerous position.

Unsurprisingly, they suffered many casualties, but enough of them made it across for sheer numbers to tell in the end. The Jacobites made a disciplined fight of it but superiority in numbers and superior gunnery won the day for William. Schomberg died at the Boyne, as did Reverend George Walker, one of the heroes of the siege of Derry. It was not a drubbing: the Jacobites had acquitted themselves well and the army was able to retreat in good order, good enough to fight on at the sieges of Limerick and the Battle of Aughrim – the battle that really finished the war and finished the old Irish Catholic elite with it – the following year.

But in European terms, the Boyne was very consequential indeed. It ended James's pretensions to an Irish crown separate from England and under French protection. James made his way back to Dublin, according to some contemporary accounts in unseemly haste. It was a significant defeat for Louis XIV's ambitions.

The flight of James II after the Battle of the Boyne, 1690.

Victory of King William III at the Battle of the Boyne, Ireland, in 1690. Print by Romeyn de Hooghe.

Aftermath

From Dublin, James made his way south to Waterford and set sail for France, where he lived out the remaining 10 years of his life as a pensioner of Louis XIV. William, likewise, saw nothing to detain him further in Ireland, so he too departed, leaving what remained of the Irish theatre of war in the hands of his capable Dutch general, Godert de Ginkel. The two sieges of Limerick and the Battle of

The Treaty Stone in Limerick City. In the background is King John's Castle (left) and the tower of St Mary's Cathedral on the right.

Aughrim – which was a close-run thing – finished the war in Ireland. A treaty made at Limerick allowed the Jacobite officers, mostly Irish Catholics whose lands had been appropriated by Cromwell, to depart for the continent. There, they and their descendants – the fabled Wild Geese – distinguished themselves in military matters, in diplomacy and in commerce. To this day, their descendants can be found in France, Spain, Portugal and Austria.

The new Irish parliament, stuffed with the victorious Williamites, was in no mood to honour any provisions of the Treaty of Limerick that they could evade. It was they, rather than the London government, that welshed on the treaty. They did so for a good reason: they had come within an ace of themselves being dispossessed of the lands they had so recently acquired, which would almost certainly have been the case had James won and established himself in Ireland. He would have had too many debts to pay to the old Catholic elite and that would have meant a reversal of the Cromwellian land settlement. So Protestant Ireland had had a very close shave and was in no mood for conciliation. That set the template for much of the next century.

The war in Europe dragged on until 1697 when it ended inconclusively. France remained the team to beat, and that would continue to be the case until 1815. Ireland settled down to the long dominance of the Williamite interest – the Protestant ascendancy – which was basically the story of the 18th century. Drogheda and the Boyne Valley, so briefly in the spotlight of history, now withdrew into the happier rhythms of civic life. The town established itself as a significant port, commanding as it did the mouth of a tidal river. However, the Boyne enters the sea rather apologetically at Mornington, just to the east, so Drogheda had no commanding bay, like Dublin to the south, with which to offer a greater anchorage or relief harbours. It remained a modest but successful port.

Once the kings had departed, the dividends of peace were soon visible, not least in the built environment. Beaulieu House (see page 168), originally built just east of the town in the 1660s, was extended early in the 18th century. Townley Hall (see page 184), just to the west, is one of the more pleasing small classical Irish houses.

Mural depicting the
Battle of the Boyne,
Belfast.

Millmount Martello Tower

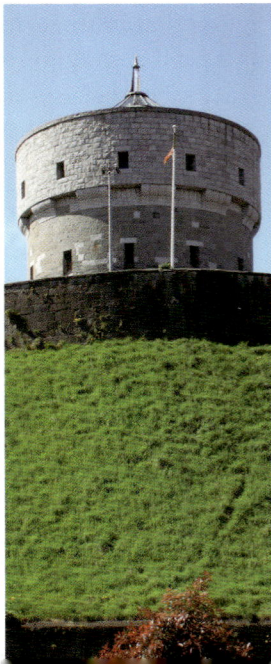

The martello tower on Millmount Hill.

Millmount, home to the Drogheda Museum, was built within a fort that was constructed on a mound, and, as a result, is clearly visible from all parts of the town of Drogheda. It is one of the most significant heritage sites in County Louth and the museum is regarded as one of the finest in Ireland.

The mound has been part of the history of the area for millennia. Although it has never been excavated, it is believed, given the proliferation of passage tombs in the Boyne Valley area, that it might be a Neolithic grave. The Celtic poet Amargin, credited as the composer of the first Irish poem, is said to be buried there.

Before the town was built, the view from the top of the mound over the surrounding countryside made it an ideal defensive point. After Hugh de Lacy had been granted the Kingdom of Meath (a larger area than the county of Meath today) he built a motte and bailey on the mound. This was later replaced by a more substantial stone fort.

The fort served as an army barracks, Richmond Barracks, throughout much of the 18th century. A variety of housing for military personnel was built on the site at this time, much of it now refurbished.

In 1808 the duke of Richmond, Charles Lennox, ordered the fortifications to be demolished to make way for the Martello tower. In the early years of the 19th century about 50 of these towers were built around Ireland, mainly on the east

Charles Lennox, fourth duke of Richmond.

coast, as a precaution against invasion by Napoleon Bonaparte.

The most recent event associated with Millmount took place in the 1920s during the Civil War when it was occupied by the IRA. The Free State Army set up a field gun and shelled the tower for several hours, causing extensive damage and forcing the IRA into retreat. In 1956, two IRA mines were discovered in a nearby street and were taken to Dublin to be destroyed.

Millmount was restored in 2000 and repurposed as the Millmount Cultural Quarter.

View of Drogheda from Millmount.

It has several permanent exhibitions dealing with archaeology, folk life, local history, geology, industry and military history, together with craft shops, art galleries and a café. There are wonderful 360° views over the Boyne Valley from the top of the tower.

Beaulieu House and Gardens

Beaulieu (meaning 'beautiful place' and pronounced 'Bewley') House, on the banks of the River Boyne outside Drogheda, was built between 1660 and 1666 by the Governor General of Drogheda, Sir Henry Tichbourne. It was the first unfortified grand house built in Ireland. The lands and castle at Beaulieu had belonged to the Plunkett family since the arrival of the Normans in

External view of Beaulieu House in County Louth

Ireland in 1082. The present house may have replaced an earlier tower house, the usual type of Norman landowner's dwelling. Some of the original defensive measures can still be seen in the grounds.

Unusually for an Irish stately home, Beaulieu didn't undergo an external makeover in Georgian times (apart from some replacement windows), and it is thought to be one of the finest examples of Irish architecture from the Restoration period of the mid-1600s. Internally a grand staircase was installed and some decorative touches were added in the early 18th century by Sir Henry's grandson, another Henry Tichbourne. The original, much simpler staircase is still in position.

Because of the Dutch bricks that were used on the facade, it is thought that a Dutch architect may have designed the house. A Dutch landscape artist, Willem Van Der Hagen, was commissioned to carry out work in the house, including a city-scape of Drogheda over the hall fireplace, painted on panel, and a magnificent painting on the ceiling of the drawing room depicting the goddess Aurora. The house today is a treasure trove of artworks, including a series of family portraits and a collection of early 20th-century paintings.

Van Der Hagen was also commissioned to design the gardens that surround Beaulieu House and he is thought to have used them to teach horticulture to his students. The four acres of walled garden and terraces have hardly been altered since they were laid down in the 18th century. A short avenue of lime trees leads to the house. The gardens are on the Boyne Valley Garden Trail and are at their most beautiful in the summer.

Beaulieu has been owned by the Tichbourne family since they came into possession of it in the 17th century. Almost uniquely, it has been passed down from generation to generation through the female line. On 27 September 2006, a 10th-generation descendant of the first Sir Henry Tichbourne, Gabriel de Freitas, a successful racing driver under the name Gabriel Konig, opened a motor racing museum in the grounds of Beaulieu. It houses a selection of her own racing memorabilia and a collection of classic racing and rally cars.

Guided tours of the house, grounds and racing museum are available on certain dates in the summer.

The walled gardens at Beaulieu in summer.

Monasterboice

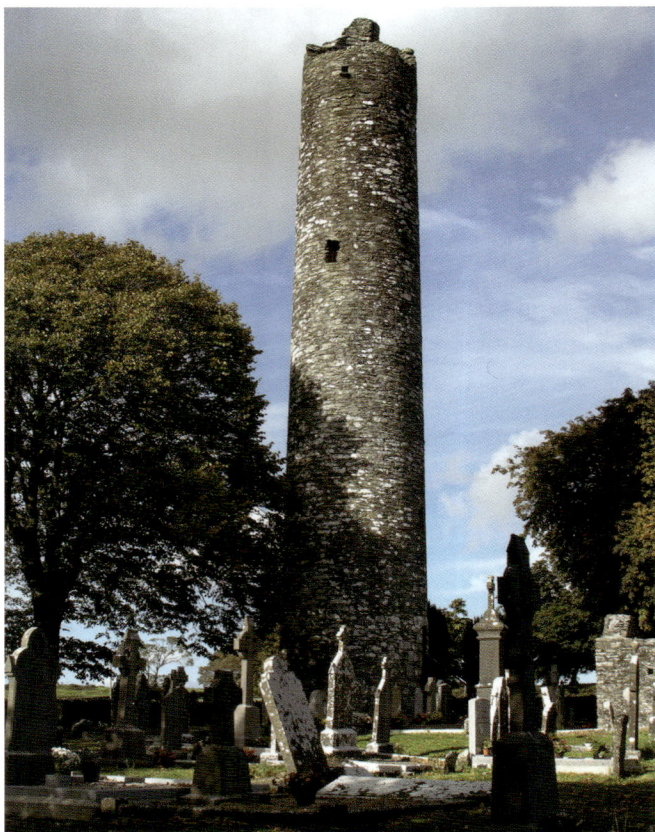

The round tower at Monasterboice.

Monasterboice, or Manistir Bhúithe, was founded by St Búithe in the sixth century. All that remains today is the 35-metre-high round tower and three high crosses. One of these is the iconic Muiredach's Cross, which stands at 5.2 metres. With its wonderful carvings of characters from the Old and New Testaments, it is regarded as the finest high cross in Ireland. It is called Muiredach's Cross because a prayer for someone called Muiredach is inscribed at the base of its west side – it is believed that this refers either to Muiredach Mac Domhnaill, an abbot of Monasterboice, who died in 923, or to Muiredach Mac

Muiredach's Cross.

The West Cross.

Cathail, the king who gifted the lands on which the abbey was built.

At 7 metres high, the West Cross is the tallest in Ireland, and is lavishly decorated. However, unlike Muiredach's Cross, the base of which is decorated with visual scriptural references and some wonderful representations of cats, the plinth is unadorned.

The very plain North Cross stands in an enclosure near the northern boundary. The shaft was damaged and replaced, and only the upper half is original. To the right of the cross there is an unusual sundial, used to mark off the hours at which the monks prayed.

The round tower at Monasterboice contained the monastery's library and other treasures. The monks would have hoped to keep them safe here in the event of a Viking attack. However, a fire in the tower in 1097 destroyed the entire contents of the tower.

Monasterboice was abandoned after the foundation of Mellifont (see page 176) in 1142. The remains of a 13th-century parish church that was built on the site of the monastery can still be seen today.

Monasterboice's round tower, church ruin and graveyard.

The abbey at Mellifont in the Mattock valley was the first Cistercian monastic foundation in Ireland and went on to become one of the wealthiest in the country. It was founded in 1142 by St Malachy (Maelmhadhóg) Ó Morgair, bishop of Down and archbishop of Armagh, after the order received a grant of land around Knowth. Donnachadh Ua Cearbhaill, the king of Oriel (Louth), provided the land and building materials, and Muircheartach Ó Lochlainn, the king of Ulster, donated cattle and gold to the foundation.

Malachy had stayed in St Bernard's Cistercian monastery at Clairvaux in France on his way to

Rome in 1139, and had been so impressed by it that he asked Pope Innocent II for permission to resign his bishopric and enter Clairvaux as a Cistercian novice. The pope refused, but when Malachy established his own monastery at Mellifont he modelled it on the French foundation, introducing the style of architecture that he had encountered at Clairvaux. He also brought several of the French monks with him to Ireland, but they squabbled with the Irish monks and returned to France.

Malachy died at Clairvaux in 1148 while he was travelling to Rome, so he didn't live to see the solemn consecration of Mellifont, which was celebrated in 1157. The high king of Ireland, Muircheartach Mac Lochlainn, was in attendance, and he presented gifts of gold, livestock and more land to the abbey. Relics of Malachy were sent from his burial place at Clairvaux to Mellifont in 1194.

Throughout his life Malachy had worked tirelessly for the reform of the Irish Church. Although he didn't live to see it, the Synod of Kells, summoned to

St Malachy, archbishop of Armagh and founder of Mellifont Abbey.

Ruins of Mellifont Abbey.

further that aim, was held at the new abbey in 1152.

By 1170, Mellifont Abbey had a population of 100 monks and 300 lay brothers. It became the model for other Cistercian abbeys and was the mother house of seven Cistercian abbeys in Ireland, including nearby Bective.

The Normans were supportive of the Cistercian foundations in Ireland, and in 1228, by order of the English king, Henry III, control of the monastery was assumed

King Henry III, Queen Eleanor and courtiers, 1240 AD. Hand coloured engraving, 1814.

by Cistercians from England, and the community there became Anglo-Norman rather than Irish, to a greater degree supplanting the monastery's Irish population. In 1380 a decree was issued forbidding the professing of Irishmen as clerics at the abbey.

By the beginning of the 15th century the Mellifont estate had increased to 48,000 acres. A 15th-century abbot almost brought about the ruin of the abbey by leasing out its lands too cheaply, thus depriving the monks of essential income.

The lavabo.

In 1539, under the legislation for the dissolution of the monasteries in Ireland, the abbey and its possessions were surrendered to the Lord Chancellor, and the buildings were leased to Lawrence Townley. In 1566, they were leased to Edward Moore on the understanding that he would turn the abbey into a manor house. In 1603, Moore's descendant, Garrett Moore, a close friend of Hugh O'Neill, offered Mellifont as a neutral place where the Irish lord could make his submission to the Lord Deputy at the end of the Nine Years' War.

In 1661, Henry Moore (after whom two Dublin streets are named), the first earl of Drogheda, took up residence in the abbey. In 1690, he hosted officers in King William's army on the eve of the Battle of the Boyne. In 1727, the abbey was bought by the Balfours of Townley Hall.

Although much of the abbey is no longer standing, it is still possible to see the shapes of the once substantial buildings on the site. The remains of the chapter house, the impressive gateway and the wonderful octagonal lavabo where the monks washed their hands before meals, make this site worth a visit. There are also the remains of some small dwellings inhabited by the poor after the abbey was abandoned.

The visitor centre has an exhibition of carvings excavated at Mellifont, including some excellent examples of masonry from the Middle Ages. Tours of the site are available from the end of May to the beginning of September each year.

In 1938 Cistercian monks founded a new Mellifont Abbey on lands that had belonged to the original foundation. This new abbey is an active one and is not open to the public.

Mellifont Abbey in a postcard from 1914.

Townley Hall

This elegant square mansion, built in 1799 on a curve of the River Boyne close to the site of the Battle of the Boyne, is a stunning example of Georgian neo-classical architecture. Designed by the leading Irish architect of the day, Francis Johnston (who also designed the GPO in Dublin and the courthouse in Kells), the house is unusual in that it has had very few alterations throughout its history.

The house was commissioned in 1794 by Blayney Townley Balfour MP, whose family took possession

of the Townley estate at the time of Cromwell. In 1957 the house and 850 acres of the estate were sold to Trinity College Dublin, which used it as an agricultural school. In 1967 a Trinity professor bought the house, and most of the land was sold off in 1969 to the Land Commission and the Forestry Commission. Townley Hall Wood was planted around 200 years ago and is open to the public under the administration of Coillte. The house, set in 60 acres of rolling parkland, is now owned by the School of Philosophy and Economic Science.

The house was built in a very peaceful location, at the end of a long, tree-lined avenue. With its beautiful interior, satisfying proportions and high-quality materials and craftsmanship (all local), it is considered to be an architectural jewel in the classical style, and is certainly Francis Johnston's masterpiece. Its most beautiful feature is

the hallway, with its elaborate stucco work, elegant spiral staircase and coffered rotunda.

There is a Neolithic passage tomb in the grounds of Townley Hall, which dates from the same era as the great mounds at Newgrange, Knowth and Dowth. The tomb is simpler than those in the Brú na Bóinne complex, and the passage leads to a single rectangular chamber. The alignment of the tomb is towards the summer solstice sunrise, symbolic of fertility and growth. A settlement at the site of the tomb was abandoned shortly after the tomb was built, leading archaeologists to believe that those buried in the tomb were of such importance that their deaths represented the demise of the community that lived there.

Townley Hall Neolithic Passage Tomb, from above.

Townley Hall has an association with the poet Francis Ledwidge, who lived nearby and liked to wander in the woods there. An early poem, 'The Robin in Townley Hall', set in those woods, tells of the joy he experienced there.

Francis Ledwidge, Irish war poet.

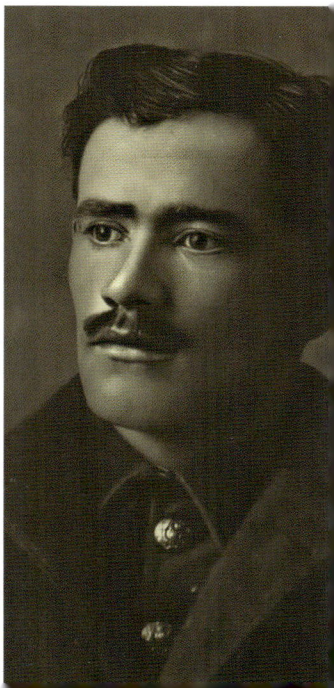

Through lovely woods and sparkling streams
I've wandered many a time.
I love to scan those distant days
And sadly I recall
Each hour I sat near the red-breasted bird
That piped in Townley Hall.

No bird the more could
cheer my heart
I heard them each and all,
Than the robin in the thorn
In the woods of Townley Hall.

The house is open to the public by appointment only.

The Francis Ledwidge Museum

On 19 August 1887 Francis Ledwidge was born in Janeville, just outside the village of Slane. He was the eighth of Patrick and Anne Ledwidge's nine children. Patrick, a farm labourer, died when Francis was five, and Anne and her children all had to work to support the family. Francis left school when he was 13 and his first poems were published in the local newspaper when he was 14. He undertook many different kinds of work, including labouring on the roads, working as a shop assistant and in copper mining. He was appointed secretary of the Meath Labour Union in 1913.

In 1912 Francis sent some of his poems to the writer Edward Plunkett, 18th Baron Dunsany, who befriended the young poet and became his patron. He showed Francis's poetry to his own literary circles in Dublin, introduced him to W. B. Yeats, and provided him with moral and financial support, allowing him to work in the library at Dunsany Castle (see page 238).

Francis Ledwidge was a patriot and nationalist and in 1914 he founded a local branch of the Irish Volunteers – this force was established to counter the Ulster Volunteers, set up to thwart the progress of the movement to introduce Home Rule. When the First World War broke out in August 1914, Francis enlisted in Lord Dunsany's regiment, the Fifth Batallion Royal Inniskilling Fusiliers, despite Dunsany's exhortations to the contrary (he even offered Francis an allowance in order to persuade him to remain in Ireland). Francis said that he felt unable to stand by while others fought 'an enemy common to our civilisation'.

Francis was killed at the third battle of Ypres in Belgium on 31 July 1917, a few days shy of his 30th birthday. A contemporary, John Drinkwater, said that 'to those who

The cottage in which Francis Ledwidge was born. It now houses the commemorative Ledwidge Cottage Museum.

know what poetry is, the untimely death of a man like Ledwidge is nothing but calamity'. Francis Ledwidge is commemorated by a plaque on Slane bridge and a memorial on the spot in Belgium where he lost his life.

Lord Dunsany was instrumental in the publication of several collections of his poetry, the first of which, *Songs of the Fields*, appeared in 1915 while he was serving in the trenches. *Songs of Peace* and *Last Songs* were published posthumously, in 1918 and 1919 respectively.

In 1981 Francis Ledwidge's birthplace was bought and restored by the Francis Ledwidge Museum Committee, and it was opened as a museum in 1982. It is considered to be a perfect example of a 19th-century farm labourer's cottage and houses a collection of the poet's works and some period memorabilia. The beautiful garden behind the museum is on the Boyne Valley Garden trail.

Francis Ledwidge's verse was inspired by the beautiful Meath countryside in which he lived and worked. His most famous poem is his 'Lament for Thomas McDonagh', written in memory of his friend, who was executed as one of the leaders of the 1916 Rising.

He shall not hear the bittern cry
In the wild sky, where he is lain,
Nor voices of the sweeter birds
Above the wailing of the rain.
Nor shall he know when loud
    March lows
Thro' slanting snows her fanfare shrill,
Blowing to flame the golden cup
Of many an upset daffodil.
But when the dark cow leaves the moor,
And pastures poor with greedy weeds,
Perhaps he'll hear her low at morn
Lifting her horn in pleasant meads.

Francis Ledwidge memorial plaque at Boesinge, Belgium, inscribed with the words of his poem 'Lament for Thomas McDonagh'.

The Hill of Slane

The Hill of Slane, which rises 158 metres above the surrounding countryside, legendary burial place of Sláine, king of the mythical Fir Bolg, is steeped in the history of Irish Christianity. At Easter 433, St Patrick is said to have lit the Paschal fire on the Hill of Slane, 16 kilometres from the Hill of Tara (although archaeologists now think that this could have happened anywhere in the Boyne Valley that was visible from the Hill of Tara), to symbolise the triumph of the light of Christ over the darkness of the druids and paganism. This angered High King Laoghaire, who presided over the lighting of the Bealtaine fire on Tara at the spring equinox each year. However, Patrick's preaching of Christianity was so convincing that, although he failed to convince Laoghaire, one of the king's chief druids, Erc, converted to Christianity and travelled the country preaching the Gospel.

A statue of St Patrick, patron saint of Ireland, stands on the Hill of Slane.

The association with St Patrick meant that the Hill of Slane was an important centre of Christianity for many centuries after his death. The ruins of the 15th-century medieval Church of St Patrick, itself a rebuild of an earlier church from the medieval era, sit on top of the Hill of Slane. The site is also home to the ruins of a college built by the Fleming family, barons of Slane from the 11th century until the 17th century – their coat of arms can be seen on the west wall of the quadrangle. It must have been an important foundation, as there are references in the Annals of the Four Masters to the presence of abbots and bishops at the monastery.

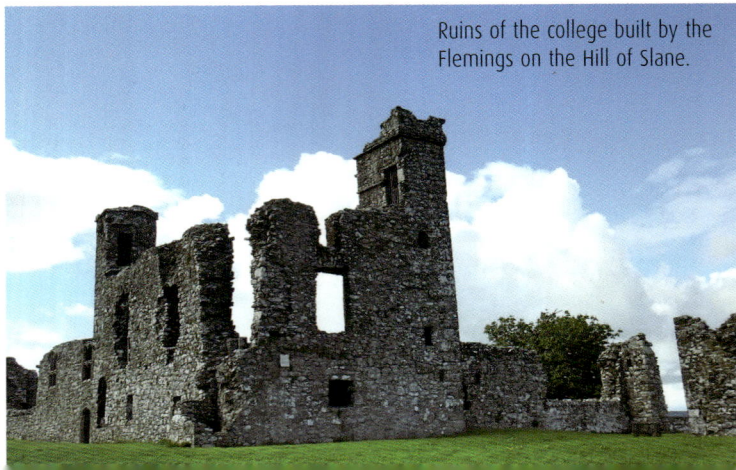

Ruins of the college built by the Flemings on the Hill of Slane.

To the west of the college are the remains of a motte and bailey constructed in the 1170s as the baronial seat of the Flemings. They later moved to a castle built on the banks of the River Boyne where Slane Castle is today.

Like so many Irish monastic foundations, St Patrick's had a chequered history. In 1542 the monastery at Slane was dissolved and its land and possessions were seized by the crown. In 1631 the Flemings restored the monastery and reinstalled the monks. Twenty years later they were driven out by Oliver Cromwell and the monastery gradually fell into a state of disrepair. The church continued to serve as the local parish church until the 18th century.

The best-preserved part of the church is the Gothic tower. The church was built within a walled enclosure on an earlier monastic foundation of St Erc, and two gable-shaped slabs within the church enclosure, referred to as the Bishop's Tomb, are thought to be all that remains of St Erc's burial place. At one time, the coffins of those whose funerals were held at St Patrick's graveyard were carried around the Bishop's Tomb three times before being interred.

Slane Castle

Slane Castle was built in a very dramatic situation, perched on a rock overlooking the River Boyne, close to the site of the Battle of the Boyne (see page 144). The Anglo-Norman Flemings, the Lords of Slane, had a fortress on the site since the 12th century. However, as staunch Catholics they were among the leaders of the 1641 rebellion and, in its aftermath, their property was confiscated by the crown. Their estates were later restored by King James II, but after his supporters were defeated at the Battle of the Boyne and the Battle of Aughrim in the Williamite War the Flemings lost Slane and its lands for ever.

By 1703 Slane had been transferred to the ownership of the Conyngham family, Scottish Protestants who had settled in Donegal in 1611, during the Plantation of Ulster. They soon owned extensive landholdings in the Donegal area. They moved their family seat to Slane soon after taking possession of the estate and built a house on the foundations of the old Fleming castle. This house, embellished with many towers of different sizes and shapes, looked nothing like the castle does today. The power and influence of the Conynghams increased as the century progressed, and by the 1780s they had decided to turn Slane into a grand castle befitting their position in society.

In 1785 they refashioned the exterior of the castle to incorporate the towers, and added battlements to the roof. The reconstruction was carried out by noted architects James Gandon, James Wyatt and Francis Johnston. The gardens were landscaped by renowned garden designer Capability Brown. The remains of a Franciscan foundation dedicated to St Patrick's disciple, St Erc, can be found in the castle grounds. This is believed to date from the 13th century, if not earlier.

In 1787 the estate was inherited by Henry Conyngham, who had been created a viscount in 1779. In 1794

James Wyatt, one of Slane's main architects, with his granddaughter Mary, painted by Sir John Everett Milais.

he married the wealthy but unconnected Elizabeth Denison, who became the mistress of King George IV and Conyngham was further elevated as first marquess of Slane in 1816. The king visited Dublin in 1821 and it was said that he was in such a hurry to get to Slane to visit his mistress that he had a straight road built along the route from the city to the castle!

In 1991 disaster struck when the eastern portion of the castle was badly damaged by fire. After a 10-year restoration project it opened to the public once again, together with the whiskey distillery that is run on the estate. Slane is probably most famous today as a venue for open-air summer rock concerts. The U2 album *Unforgettable Fire* was recorded there in 1984.

Jon Bon Jovi and Lord Henry Mountcharles, owner of Slane Castle, announcing Bon Jovi's concert at the venue.

Ardmulchan Church

Little remains of Ardmulchan Church apart from the square bell tower, the ground floor of which is still almost completely intact. The tower was built in the 13th or 14th century, and the ruins of the church probably date from the 15th century. The church was suppressed in 1613.

A motte, church and several smaller chapels were built on pre-Norman foundations at some time in the 12th century. Ardmulchan was de Lacy land, and defending this vulnerable stretch of the River Boyne was an imperative in the early days of the Norman occupation. The church and chapels were granted by charter to the monks of the Abbey of St Thomas the Martyr in Dublin. Although the Ardmulchan

remains are scant, it is the location of the church that makes it a recommended stopping place along the Boyne Valley Drive route. Built in an elevated position on the south bank of the River Boyne, it provides an ideal spot from which to drink in some breathtaking views. William Wilde, who was given to extolling the charms of the area, wrote that: 'Here the true beauty of the Boyne, its real rhyme-like character, commences.'

Looking out over the landscape, you will see the round tower at Donaghmore (see page 204) and the ruins of Dunmoe Castle in their lush green setting of gently sloping farmland. If you take a walk along the banks of the river you'll catch a glimpse of the impressive Scottish-style 'castle', Ardmulchan House, built in the early 20th century.

A view of the Boyne Valley
from Ardmulchan in autumn.

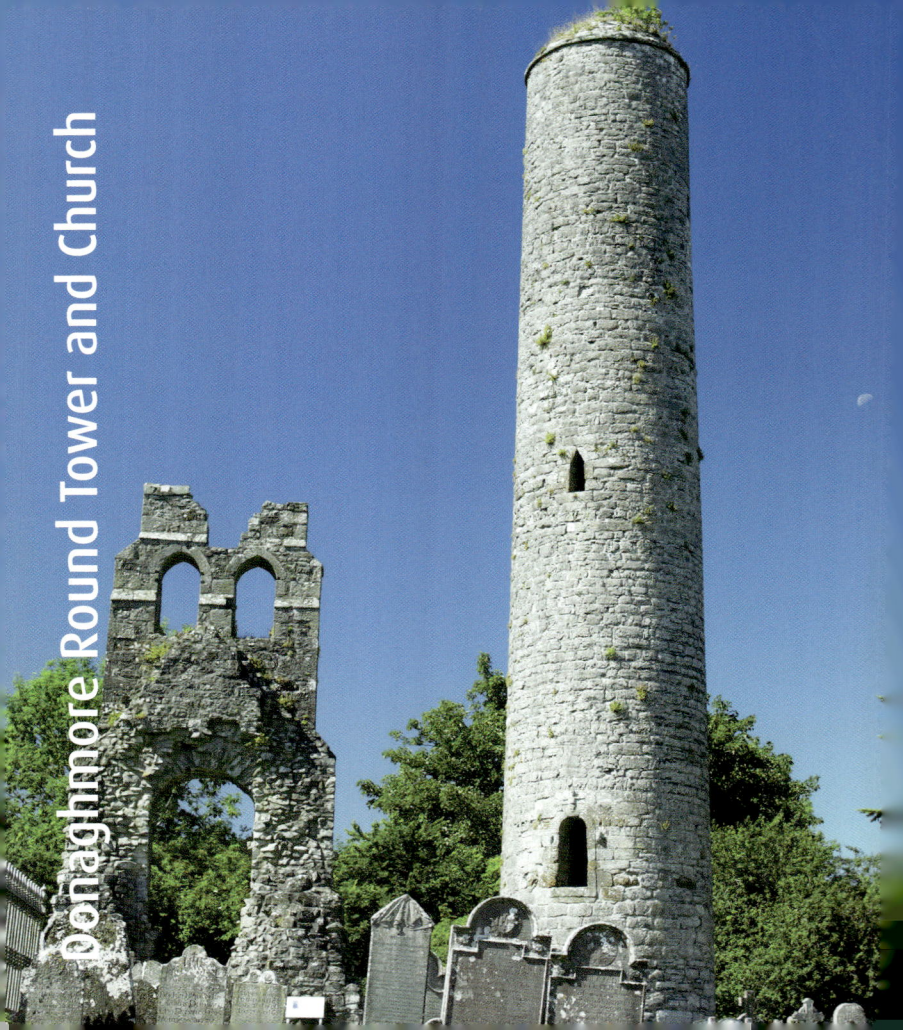

Donaghmore Round Tower and Church

One of the best-preserved round towers in Ireland, Donaghmore stands more than 30 metres tall, with a base circumference of just over 20 metres. It is unique in that it has a carving of the crucifixion above the entrance, with a stone head on either side of it.

The remains of the tower and church at Donaghmore.

Although little remains of the church, Donaghmore (from Domhnach Mór, meaning the Great Church) was an important site from the beginnings of Christianity. St Patrick himself is said to have blessed the original church there, before giving it into the care of one of his followers, St Cassanus.

The upper part of the tower was badly restored in 1841 and the top does not conform to the usual pattern.

The site of original monastery foundation at Donaghmore is mentioned in the deeds that were written into the blank pages of the Book of Kells (see page 210). At some point before the end of the 10th century the monks at Kells bought the land at Donaghmore for 20 ounces of gold.

The monastery would have had a celibate population of monks in its early years, but this later evolved into a community of married clergy and lay monks. Like other monastic foundations on the Boyne, Donaghmore's proximity to the river made it vulnerable to Viking raids, and it was attacked and plundered on several occasions.

The original church on the site was replaced, probably in the 13th century, and the present ruins date to the 15th century. This later church incorporated Romanesque carvings of heads, probably taken from the 13th-century structure.

The graveyard is also interesting, with an early Christian cross and several 18th-century gravestones. There is also the grave of an unnamed Croppy, one of the participants in the 1798 Rebellion.

Kells Monastic Site

Long before Christianity arrived in Ireland, Kells (from the Irish An Ceanannas Mór, meaning the 'great residence or fort') in County Meath was a significant settlement. It had royal connections as the residence of King Conn of the Hundred Battles and King Cormac Mac Airt.

Kells rose to prominence in the sixth century when High King Diarmuid Mac Cearúil granted St Columba the fort and surrounding land in around 550. Columba founded a monastic settlement there. However, in 563, having been responsible for a bloody battle in a dispute over the intellectual copyright in a psalter, he left Ireland with several monks and went into self-imposed exile on the island of Iona off the west coast of Scotland. He founded a famous monastery there that flourished for centuries. Columba never returned to Ireland, but after several Viking invasions of Iona in the eighth and ninth centuries, the monks there fled to Ireland where they were given a safe haven at Kells.

The Iona monks refounded their monastery at Kells, building a new abbey there on the site of an ancient hill fort. The building work lasted seven years. The abbey church of the new foundation was consecrated in 814, when the abbot of Iona, Ceallach, moved to Kells. In 878, the remains of Iona's founder, St Columba, were moved to Kells in a reliquary shrine, and Kells became the principal Columban foundation.

The abbey's most famous treasure, the Book of Kells, may simply have been kept there, having been created elsewhere. Whether it travelled to Kells with the reliquary, or was created at Kells or at some other

St Columba's House, Kells.

location or locations in Ireland, has long been the subject of speculation. However, this wonderful illuminated manuscript of the four gospels, one of the world's oldest books, was kept at Kells for several centuries.

Kells was raided by Vikings three times during the 10th century, and the following century saw many incursions by the native Irish. The Book of Kells was stolen in 1006, but was later returned, minus its glorious jewelled cover. It was kept at the abbey throughout the medieval period, and in the 12th century details of some of the abbey's land transactions were recorded on its blank pages.

Although the abbey was dissolved in the 12th century and the abbey church became the parish church of Kells, the Book of Kells continued to be kept there. In the 1650s Cromwell's troops were billeted in the town of Kells and the Book of Kells was removed from the abbey for safekeeping and sent to Dublin. It was donated to Trinity College Dublin in 1661 and has remained there ever since. It is kept in the Long Room of the college library and can be viewed by the public. You can see a replica in the Kells Heritage Centre.

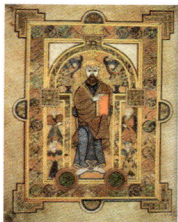

Page from the Book of Kells.

Cross of St Patrick and St Columba, the South Cross, with carved biblical scenes.

Today only part of the monastery remains and the town of Kells was built around several of the monastic structures. The remains of the circular monastic huts now surround several of the monastic treasures that are located in the town. The 10th-century 30-metre high round tower, built to protect the monks and the abbey treasures from the invaders, is one of these. The tower is a little out of the ordinary in that it has five windows at the top, rather than the usual four. It is believed that these were meant to provide vantage points over the five roads that led into Kells.

Close to the round tower is one of several of the high crosses for which Kells is famous. This is the ninth-century South Cross, one of the oldest of the crosses.

The High Cross, also dating from the ninth century, stands outside Kells Courthouse. Its carvings depict stories from the Bible, and the sculptors also included

The Long Room, Trinity College Library, Dublin, which houses the Book of Kells.

a lot of decorative flourishes, including spirals, which are believed to have been inspired by the carvings at the Brú na Bóinne complex. Within the walls of the monastic site are the North Cross, again with depictions of biblical stories, and the so-called Unfinished Cross.

A small oratory with a steep pitched stone roof, St Columba's House, in Church Lane, dates from the 10th century. This is the supposed location of the writing of the Book of Kells.

The Anglican church on the site was built in 1778. The round tower and the high crosses are designated national monuments, and the site is open to the public.

The Unfinished Cross and round tower.

The Spire of Lloyd

Three kilometres to the north-west of Kells in County Meath and 40 kilometres inland, is one of Ireland's strangest follies. At the top of the Hill of Lloyd, overlooking the River Blackwater, stands a tower, which, at first glance, looks exactly like a lighthouse. It is styled as a 30-metre high Doric column with a glazed lantern at the top, and has been referred to as 'Ireland's only inland lighthouse'. The views from the tower are spectacular. Internally, there is a 164-step spiral staircase leading to the glassed-in gallery at the top, from which the views of the countryside are breathtaking – on a clear day you can see as far as the Mourne Mountains in County Down, Northern Ireland.

As a lighthouse, the location of the Spire of Lloyd, at a distance of 40 kilometres from the sea seems strange. Why was it built?

In the late 18th century the fashion for building follies (structures built for decoration rather than function) was at its height. The Spire of Lloyd was built in 1791 by Thomas Taylour, first earl of Bective, Lord Headfort, in memory of his father (also Thomas Taylour). The purpose of the tower is unknown, but it may have been built to provide employment for the people of the area during a local famine, making it one of Ireland's so-called 'famine follies'. Some people have speculated that it was built so that the Headfort family could watch horse-racing from the top (although it predates the Kells racecourse by 50 years), or so that they could view their ships coming into Carlingford Bay, 40 kilometres away. It was designed by the architect of King's Inns in Henrietta Street, Dublin, Henry Aaron Baker, who was a pupil of James Gandon. The tower bears an inscription that reads:

This pillar was designed by Henry Aaron Baker Esq. architect, was executed by Mr. Joseph Beck stone cutter, Owen Cabe head mason, Mr. Bartle Reilly overseer Anno 1791.

On the east side of the tower there is a plaque depicting the Headfort family's coat of arms, together with their motto – *Consequitur quodquinque petit* (He follows what he seeks).

Recent investigations show that the tower is actually built on top of an Iron Age ring fort, the origins of which may go back to the Bronze Age.

The name of the spire comes from the original name of the hill, Mullach Aití, anglicised as Mulloyde and thence to Lloyd. Mullach Aití provided a lookout point for the Kingdom of Midhe (Meath) against incursions from the Kingdom of Breifne (Cavan). The elevated site made a good royal encampment – in the Táin it is reported that Queen Maeve camped here with her army on her way to Ulster, and Edward the Bruce of Scotland set up camp after his victory at Kells in 1314, during his invasion of Ireland.

The land around the tower is now a community park, and is the starting point for the Ringfort and Blackwater River loop walk. The spire is open to visitors on bank holiday Mondays.

Sunrise over the River Blackwater.

The Loughcrew Cairns

Loughcrew is one of the four most important passage tomb sites in the country (Brú na Bóinne, Carrowmore and Carrowkeel in Sligo are the others). This collection of around 25 Neolithic passage tombs, also known as the Hag's Hills or the Hills of the Witch (Sliabh na Caillighe), near Oldcastle, County Meath, dates from around 3300 BCE, contemporaneous with Newgrange and predating Carrowkeel and Carrowmore. The site is large, spread over the hilltops of Carnbane East, Carnbane West, Carrickbrack and Patrickstown Hill. As at Knowth, there are smaller satellite tombs clustered around the larger cairns. There is a legend that the cairns were formed when the Hag of Beara came to the area to perform a magical feat. She filled her apron with stones and, leaping from hilltop to hilltop, she dropped a stone from her apron on each hill. Having positioned three stones successfully, she slipped when attempting to place one final one at Patrickstown, fell down the hill and broke her neck.

One day in 1865, a schools inspector, Eugene Conwell, went for a picnic with his wife in the hills at Loughcrew. He discovered the cairns and spent many weeks exploring them. He named the cairns, using the letters of the alphabet. The largest is known as Cairn T, and, as at Newgrange and Knowth, has a corbelled roof and cruciform chamber. One of the kerbstones of this cairn has a depression known as the 'Hag's Chair'. It is said that the Hag of Beara used to sit on this rock to watch the stars. In 1980 it was discovered that, as at Knowth, the solar alignment of Cairn T is towards the spring and autumn equinoxes.

A cairn at Loughcrew.

It is thought that the carvings on the stones inside the chambers at Loughcrew predate those at Brú na Bóinne, giving rise to the idea that the carved stones at Newgrange, Knowth and Dowth were found elsewhere and repurposed.

At the time of the Penal Laws in the 18th century, Loughcrew provided a meeting point for Catholics, and a cross carved into the seat of the Hag's Chair has given rise to the belief that it was used as a mass rock for the clandestine celebration of mass.

The Loughcrew Hills are the highest in County Meath and there is a spectacular viewing point at the top of Patrick's Hill that rewards the hiker with a panorama stretching from the Wicklow Mountains in the south to the Mourne Mountains in County Down to the north.

View from Carnbane East.

Loughcrew House.

Loughcrew House and Gardens

The Loughcrew demesne, situated below the Loughcrew cairns, was the family seat of the Anglo-Norman Plunkett family, the most famous member of which was the martyred St Oliver Plunkett. For their role in the Irish Confederacy of the 1640s, the Plunkett family was dispossessed in 1652, and their estate was given to the Napier family in 1655. They expanded it until it covered an area of 180,000 acres. The house burned down several times, but has been restored by the current Napier owners. The wonderful gardens, created in the 17th and 19th centuries across an area of around six acres, are open to the public.

Trim Castle

Trim Castle.

Looking at the majestic edifice of Trim Castle, it is difficult to imagine its simple origins – a wooden fort protected by a ditch and stockade, erected by the Norman knight Hugh de Lacy. Sir Hugh began to build the stone castle that replaced it in 1176.

In 1172 Henry II gave the Liberty of Meath (a much larger area than the present-day county of Meath, extending as far west as the River Shannon) to de Lacy. It was an attempt to push back against a land grab by Strongbow (Richard de Clare, second earl of Pembroke) – Henry was worried that it would ultimately become a power grab, with Strongbow declaring himself king of Ireland. The construction took more than 30 years, and de Lacy died before it was finished. His heir, his son Walter, brought the enormous project to its conclusion. It must have been worth the wait – Trim was the largest Norman castle in the country, with high defensive walls enclosing more than three acres of ground.

The huge three-storey castle, built on a crossing of the River Boyne on the very edge of the Pale, marking its northern boundary, is still a looming presence in the town of Trim. The keep is a 20-sided cruciform tower, buttressed by four square towers, three of which are still standing. It was surrounded by a moat, and the whole structure was protected by a curtain wall and a barbican. Defensive towers were built into the wall at regular intervals, and additional protection was provided on one side of the castle by the swift waters of the river. It was a luxurious building for the time, with glassed windows and a tiled roof. The masonry was covered in a protective white render, as was usual at the time. The castle must have dominated the surrounding landscape, particularly on sunny days.

In its heyday, Trim, like many large castles, was self-sufficient, almost like a small town. It had a chapel, stables and

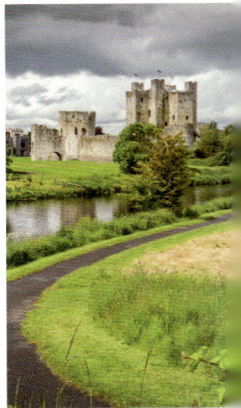

Trim Castle from the banks of the River Boyne.

The walls of Trim Castle were used in the 'Siege of York' sequences in the 1995 film *Braveheart*.

even a mint. Buildings were added in every century from the 14th to the 17th. It is an indication of the significance of the castle that no fewer than seven parliaments were held there in the 15th century. It declined in importance from the 16th century, and the fabric of the buildings began to deteriorate, but it was refortified in the mid-17th century during the Irish Confederate Wars. In 1649 it was captured and occupied by Cromwellian forces after the sacking of Drogheda, and the entire garrison fled.

In the 19th century the castle was bought by the Dunsany Plunkett family. After they sold it to the state in 1993 a programme of conservation was carried out by the Office of Public Works and the castle was reopened to the public in 2000. In 1995 it was used as a location for the film *Braveheart*, a proxy for the cities of York and London.

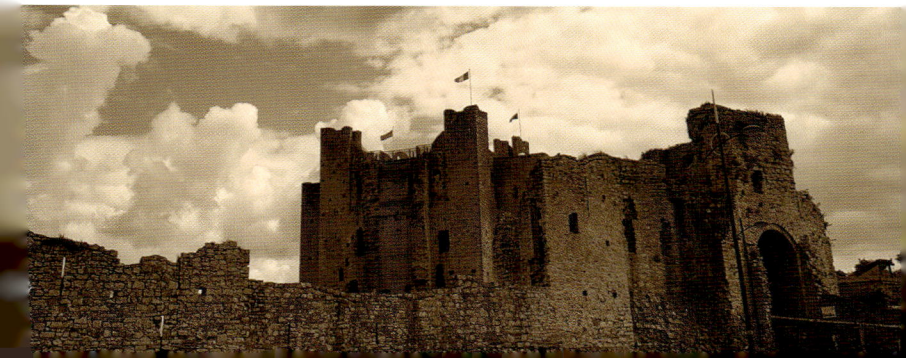

Bective Abbey

Bective Abbey was established in 1147, five years after the foundation of the Cistercian mother house at Mellifont. King Murchad O'Maeil-Sheachlainn of Meath, an important patron of monastic foundations, established Bective in a beautiful site overlooking the River Boyne less than 10 kilometres from the site of the town of Trim. Nothing remains of the original abbey, which was absorbed into later structures in the 13th, 15th and 16th centuries. The 15th-century cloister is particularly well preserved, its Gothic arches an exemplar of restrained Cistercian architecture. A beautifully carved pillar depicts a figure with a crozier, believed to be the abbot who ordered the building of the cloister.

Bective Abbey

In its heyday Bective was one of the most important abbeys in Ireland, as evidenced by the fact that its abbot sat in the Parliament of the Pale. It was the burial place of the body of the Lord of Meath, Hugh de Lacy, after his murder in 1195 – his head was sent to St Thomas's Church in Dublin. In 1205 the bishop of Meath intervened in a dispute between the two churches and had de Lacy's body reunited with his head at St Thomas's.

Each century of the abbey's existence as a monastic foundation witnessed at least one dramatic event. In 1227 the abbot was imprisoned for his part in riots at Jerpoint Abbey in County Kilkenny. In 1380 King Richard II issued a decree prohibiting any Irishman or any enemy of the king from becoming a monk at Bective. In 1488 the abbot of Bective was involved in the Lambert Simnel rebellion (Simnel was a pretender to the English throne), but was pardoned for his part in the affair.

In the 15th century the number of monks at Bective began to decline, and some of the abbey buildings were demolished in an attempt to adapt to the smaller population. A new western facade was protected by a fortified tower, and another tower was built at the south-western corner of the cloisters.

In 1536, during King Henry VIII's dissolution of the monasteries, the abbey buildings and 1,580-acre estate were confiscated and the abbot was pensioned off. The buildings were rented to Thomas Agard, before the estate was granted to Andrew Wyse, Vice-Treasurer of Ireland, in 1552. It subsequently changed hands several times. In the 17th century the abbey buildings were repurposed as a manor house, with the addition of a second accommodation storey and the installation of fireplaces with tall chimney stacks. This gave it the appearance of a castle and it was used as a location in the 1995 film *Braveheart*.

King Richard II setting out upon his invasion of Ireland.

The Golden Coffin

There was a story in the locality that a golden coffin had been buried at the abbey. In the late 19th century a local man, Andy Gossan, was the only person who knew of its whereabouts; he told his sister who passed on the information to the parish priest of Kilmessan. He died in 1927 without having divulged the coffin's whereabouts, and it has never been found.

Bective was bought by the state in 2012 and is managed by the Office of Public Works. The abbey is a protected monument.

The cloisters at Bective Abbey, used in the 1995 film *Braveheart*.

The Hill of Tara

The Mound of Hostages at the Hill of Tara.

Described as the 'most consecrated spot in Ireland' by the leaders of the Celtic Revival movement in the early years of the 20th century, the Hill of Tara, near Skryne in County Meath, commands a breathtaking view over the surrounding countryside – supposedly taking in a quarter of the island of Ireland. Its story is one in which myth and reality are interwoven. One of the most significant sites in Ireland, it was the inauguration place and legendary residence of the high kings of Ireland, although, according to archaeologists, no permanent residences were ever built on the hill. However, Tara was important long before the high kings. The remains of at least 20 ancient monuments have been found there – one of them, a passage tomb dating, like Newgrange, from around 3200 BCE, is known as 'the mound of the hostages' (Dumha na Giall).

The Lia Fáil (Stone of Destiny), a standing stone located within the Forrad (the Royal Seat), is one of Tara's most important monuments. It was the inauguration stone of the high kings, and it is said to have cried out whenever a rightful Irish or Scottish king placed his foot on it. In 1022 the high kings finally abandoned Tara.

With the advent of Christianity Tara once more became important. At the spring equinox it was the tradition to light a huge fire on Tara, and no other fire could be lit, on pain of death, until the Tara fire was blazing. When St Patrick lit a Paschal fire that night on the neighbouring Hill of Slane (see page 192), or on some other hill that was visible from Tara, it was in direct defiance of High King Laoghaire, who rejected Christianity. Laoghaire was watching from his vantage point on the Hill of Tara and sent his druids to vanquish Patrick. However, even the most powerful incantations and spells of the druids were not strong enough to extinguish the fire and despatch Patrick, and the fire burned on. Fresh from his triumph, Patrick soon afterwards used a shamrock plucked from the Hill of Tara to explain the meaning of the Holy Trinity to Laoghaire – it was said that the whole hill was then covered with shamrocks. This was the death blow for paganism in Ireland, as Laoghaire, although he himself didn't convert to Christianity, gave Patrick permission to preach the Gospel throughout the country.

The 1798 Rebellion

On 26 May 1798 the Hill of Tara, symbol of Irish sovereignty, was the scene of a heavy defeat by government forces of a 4,000-strong army of Meath's United Irishmen and Defenders.

The rebels retreated to the ruined church on the hill but were driven out. The government forces had few casualties, but 400 dead rebels were buried in a mass grave on the hill. The Lia Fáil was moved there to mark the spot. Two gravestones on the hill commemorate the battle.

Daniel O'Connell

In 1843 O'Connell chose Tara as the location for one of his 'Monster Meetings'. An estimated one million people gathered there to hear him speak against the Act of Union of Great Britain and Ireland.

Repeal meeting at Tara, 1843.
The Illustrated London News.

Dunsany Castle

The original Dunsany Castle was built in 1180–1200 by Hugh de Lacy, on the site of an older motte and bailey fortification. Dunsany was ancillary to de Lacy's main seat at Trim Castle.

Parts of de Lacy's castle at Dunsany are still standing and the four towers are an integral part of the present castle, which was built at different stages, mainly during the 18th and 19th centuries.

Dunsany Castle is now three times larger than the original building.

Apart from a short period of exile by Cromwell to Connacht, Dunsany has been in the continuous ownership of the Plunkett family since the early 15th century, when Sir Christopher Plunkett acquired the estate through his marriage. He became the first baron of Dunsany several decades later. The family's fortunes waxed and waned

over the centuries – John Plunkett, the second baron, supported the rebellion of the pretender Lambert Simnel in 1487; the fifth Lord Dunsany, Robert Plunkett, was accused of involvement in the Silken Thomas revolt of 1534; the 11th Lord Dunsany, Randall Plunkett, supported the cause of King James II in the Williamite Wars and was dispossessed, but was restored to his estates after the Treaty of Limerick in 1691. Horace Plunkett, younger brother of the 17th Lord Dunsany, was a keen agriculturalist who promoted the co-operative movement in Ireland. The Plunkett family still lives at

Dunsany, which is believed to be the only house in Ireland that has enjoyed such a long occupation by a single family. The castle is also one of the longest occupied homes in the country.

The Dunsany estate was greatly reduced in size when a large portion was transferred to its tenants under the Land Acts in the late 19th and early 20th centuries. However, the castle still sits in the grounds of its original demesne, which is surrounded by a drystone 'famine wall', built as a relief project during the years of the Great Famine in the mid-19th century. The demesne includes the family's private 15th-century chapel, dedicated to St Nicholas of Myra, and known locally as 'the Abbey'. The roof is gone, but there are some highly regarded medieval carvings, including those on the baptismal font and the double tomb of an early baron and his lady.

Sir Horace Plunkett (right) with Henry Wallace.

The very ancient-looking gateway to the castle is in fact a Gothic facsimile and conceals a gate lodge. It faces a late medieval stone pilgrim cross, the Dunsany Wayside Cross, erected when Dunsany was on a busy pilgrim route.

Tours of the beautiful interior of the castle, with its extensive collection of paintings, furniture and ceramics, can be taken during July and August. One of the most impressive rooms in the castle is the library, decorated in the Gothic Revival style with

Opposite the avenue leading to Dunsany Castle is the medieval Dunsany Wayside Cross.

a wonderful ceiling. The library houses a fine collection of books and the table at which the 18th Lord Dunsany, Edward John Moreton Drax Plunkett (1878–1957), writer and dramatist and collaborator of W. B. Yeats and Lady Augusta Gregory, worked with his protegé, the poet Francis Ledwidge (see page 188).

Edward Plunkett, 18th Baron Dunsany (1878–1957).

Lord Edward Dunsany and his wife,
Lady Beatrice, early 20th century.

Skryne Church

A little to the west of the legendary Hill of Tara, in the Tara/Skryne Valley, is the Hill of Skryne. At 135 metres Skryne is higher than Tara, but is little known outside the area. The well-preserved ruin of a 14th-century church sits on top of the hill, in a position with commanding views of the surrounding countryside. Those who attended mass at the church must have felt very close to God's creation.

The church supplanted a much earlier monastic foundation dating from the eighth or ninth century that was associated with St Columba/Columcille, who founded the famous monastery at Kells. It is believed that Columba's relics were brought to the monastery in 875 – the name Skryne comes from 'Scrín Cholm Cille' (Columcille's Shrine). A carving above the doorway is said to be a representation of the saint.

The monastery was plundered several times between the 10th and 12th centuries, by the Vikings and other raiders, but the monks remained. Columcille's shrine was

Doorway with carved figure of St Columcille above.

stolen in 1027, but was later recovered and returned to the monastery.

In 1173, the Anglo-Norman Lord of Meath, Hugh de Lacy, granted the Barony of Skryne to one of his favourite knights, Adam de Feypo. The fourth baron, Francis de Feypo, built an Augustinian friary and chantry on the site of the earlier foundation.

The 16th-century tombstone of the Marward family, barons of Skryne.

The church is known as Skryne Tower, for the large square tower that was added in the 15th century. Katherine de Feypo, Francis's heir, married into the Marward family, and her heirs inherited the Skryne title. There is a Marward family tomb close to the church.

At the foot of the Hill of Skryne is O'Connell's pub, which featured in the Guinness television advert released at Christmas 2011.

Opposite:
O'Connells pub, Skryne.

There are numerous places of interest in the Boyne Valley area, including some dedicated to younger visitors, those seeking active pursuits and other cultural activities. Just a few are mentioned here.

One of the attractions at Tayto Park.

Other Places of Interest

Tayto Park is a theme park and zoo sponsored by the Irish potato crisp brand, Tayto. It is located in the townland of Kilbrew, in County Meath.

Fantasia Water Park, Drogheda, is a leisure complex with waterpark, bowling, roller-skating, crazy golf, rock-climbing and more.

Red Mountain Open Farm, Donore. Young children are able to play indoors and outdoors at this attraction, close to the Newgrange Monument.

The Zone, Navan is an activity centre, offering a variety of action fun, such as karting, laser tag and archery, amongst other activities.

Highlanes Gallery, Drogheda, is a public art gallery and visual arts exhibition centre sited in the former Drogheda Franciscan Church and part of the Friary; known locally as the 'High Lane Church'.

Newgrange Farm is a family-run working farm and educational site, where children can try some hands-on experiences with farm animals.

The Irish Military War Museum at Starinagh offers an insight into Irish participants' involvement in both the First World War and the Second World War, as well as other military conflicts in world history.

Select Bibliography

Coffey, George, *Newgrange and Other Incised Tumuli in Ireland,*
Poole: The Dolphin Press, 1912. Reprinted 1977.

Harbison, Peter, *Treasures of the Boyne Valley,*
Dublin: Gill & Macmillan, 2003.

Hensey, Robert, *First Light: The Origins of Newgrange,*
Oxford and Havertown PA: Oxbow Books, 2015.

Mac Uistin, Liam, *Newgrange, Knowth & Dowth: Exploring the Majestic Passage Tombs of Ancient Ireland,*
Dublin: The O'Brien Press, 2009.

O'Kelly, Michael J., *Newgrange: Archaeology, Art and Legend,*
London: Thames and Hudson, 1982.

Wilde, William, *The Beauties of the Boyne and Blackwater,*
Dublin: McGlashan & Gill, 1849.

Standing stones at Newgrange on a foggy, freezing cold day at the winter solstice.

Picture Credits

The publisher gratefully acknowledges the following image copyright holders. All images are copyright © individual rights holders unless stated otherwise. Every effort has been made to trace copyright holders, or copyright holders not mentioned here. If there have been any errors or omissions, the publisher would be happy to rectify this in any reprint.

1 National Library of Ireland on The Commons @ Flickr
3 Marshall Ikonography, Alamy
4 Design Pics Inc., Alamy
7 Ian MItchinson, S/Stock_
10 Marshall Ikonography, Alamy
12 Wiki Creative Commons
15 Pictorial Press Ltd, Alamy
16 Chronicle Alamy
17 National Library of Ireland on The Commons @ Flickr
19 deadlyphoto.com/ Alamy
20 Design Pics Inc., Alamy
24 Adam Bialek, S/Stock
27 Teapot Press/Arianna Nicora
28 Notafly, Wiki Creative Commons
31 Design Pics Inc., Alamy
32 MN Studio, Alamy
34 MN Studio, Alamy
36 O Laparoto, S/Stock
37 Robert Harding, Alamy
38 Kerem Gogus, S/Stock
41 PA Images, Alamy
43 PA Images, Alamy
44 UnaPhoto, S/Stock
44 Douglas Pfeiffer, S/Stock
46 Jon Sullivan, S/Stock
47 PA Images, Alamy
49 Rik Hamilton, Alamy
51 Regan Buker, Creative Commons, Flickr
53 Regan Buker, Creative Commons, Flickr
54 Montagu Images, Alamy
55 Artepics, Alamy
57 Wiki Creative Commons
59 Ivy Close Images, Alamy
61 Teapot Press
63 Chronicle, Alamy
66 Spectrum Blue, S/Stock
67 Jononmac, Wiki Creative Commons
69 SOTK2011, Alamy
71 Blaine Horrocks, Creative Commons, Flickr
72 Hanging Bear Media, S/Stock
75 Teapot Press / Arianna Nicora
77 RM Ireland, Alamy
78 Framalicious, S/Stock
79 PA Images, Alamy
81 Anthony Murphy
82 Hanging Bear Media, S/Stock
84 Archaeo Images, Alamy
86 UnaPhoto, S/Stock
88 Jessica Spengler, Creative Commons, Flickr
89 Warren LeMay, Creative Commons, Flickr
91 Warren LeMay, Creative Commons, Flickr
92 Jorge Tutor, Alamy
94 Daniela, Creative Commons, Flickr
94 LM Spencer, S/Stock
95 Teapot Press / Arianna Nicora
96 Teapot Press / Arianna Nicora
99 Warren LeMay, Creative Commons, Flickr
100 Warren LeMay, Creative Commons, Flickr
101 MN Studio, S/Stock
102 Framalicious, S/Stock
104 Pecold, S/Stock
105 Hanging Bear Media, S/Stock
107 Ken Williams, Creative Commons, Flickr
108 Kanstantsin Markevich, Alamy
110 Adam Bialek, S/Stock
112 Weredragon, S/Stock
112 Teapot Press
114 Frank Bach, S/Stock
116 Frank Bach, S/Stock
117 MN Studio, S/Stock
118 Wiki Creative Commons
119 Theilr, Creative Commons, Flickr
120 National Geographic, Alamy
123 AA World Travel Library, Alamy
124 Derick Hudson, S/Stock
125 Derick Hudson, S/Stock
126 Derick Hudson, S/Stock
127 Wiki Creative Commons, enhanced
128 Archaeo Images, Alamy
130 Bill Bachman, Alamy
131 Bill Bachman, Alamy
132 Yannick Luthy, Alamy
134 Cinematograph Pictures, S/Stock
134 Roberto Rizzi, S/Stock

Abbreviation: S/Stock: Shutterstock

One-Minute
Wellness

SELF
CARE

⏳

One-Minute Wellness

SELF CARE

ABI SMITH

Michael O'Mara Books Limited

First published in Great Britain in 2026 by
Michael O'Mara Books Limited
9 Lion Yard
Tremadoc Road
London SW4 7NQ

EU representative:
Authorised Rep Compliance Ltd
Ground Floor
71 Baggot Street Lower
Dublin D02 P593, Ireland

This product is made of material from well-managed, FSC®-certified
forests and other controlled sources. The manufacturing processes
conform to the environmental regulations of the country of origin.

For further information see
www.mombooks.com/about/sustainability-climate-focus
Report any safety issues to product.safety@mombooks.com and see
www.mombooks.com/contact/product-safety

ISBN: 978-1-78929-850-5 in hardback print format
ISBN: 978-1-78929-858-1 in ebook format

1 2 3 4 5 6 7 8 9 10

Cover and design by Natasha Le Coultre, using illustrations from
Shutterstock
Typeset by Barbara Ward
Printed and bound by China

www.mombooks.com

MIX
Paper | Supporting
responsible forestry
FSC® C010256

Contents

Introduction

Self-care is the conscious act of looking after the one person who often seems to be neglected: you. Amid busy schedules, endless responsibilities and digital noise, taking the time to care for yourself has never been more important.

This book is a gentle companion for anyone seeking balance, comfort and compassion. Dip in and read any page. Any time. The entries are short – it will take about a minute to read each one. You'll find ideas and exercises to remind you that you matter.

All you need is the willingness to make a moment for yourself every day. With a small amount of attention to your self-care, you'll feel more connected to yourself and the world around you. You'll delight in small pleasures. You'll recognize that you are capable of setting healthy boundaries and connecting with positive energy.

This book is divided into the seven self-care pillars. Each plays a vital role in balancing the different areas of your well-being.

Adopt some of the practices. Embrace the advice. Above all, remind yourself that you have the most power to make a difference in your life.

Your Mind

Understanding and caring for
your mental health should be
a non-negotiable part of your daily
routine. After all, it's important
to strengthen your mind as
well as your body.

Look after your mind and give it the care it deserves

As you go about your daily routine, pay attention to your feelings. Try filtering in a few simple steps that can bring you back to yourself: focus on your plans rather than comparing yourself to others; soften your inner dialogue; accept your humanness with all its strengths and weaknesses. You don't need to change. That's the practice and that's the goal.

Picture your mind as a garden and your thoughts as seeds

You can decide what to plant in the soil: seeds of love, positivity, kindness; or seeds of negativity, shame and fear.

You might look at your friend's garden and think they have the most vibrant, luscious and sweet-smelling orchard – one that you could never hope to match. But, by focusing elsewhere, you will be neglecting your own land.

Be kind to yourself. Spend time in your garden and make it a place filled with wonder and beauty.

'A bird sitting on a tree is never afraid of the branch breaking, because its trust is not on the branch, but on its own wings.'

Charlie Wardle

Rely on yourself

Self-belief is key to mental resilience and this quote is your reminder that you should always rely on yourself. There is no *trying*, just *doing*. And you are stronger than any difficulties that come your way.

Exhausterwhelmulated

Adjective: *The feeling of being worn out, stressed, overwhelmed and overstimulated all at the same time.*

A feeling that's quite relatable, isn't it? This is your cue to stop. To exhale. To regroup. There is no challenge that can't be overcome one step, one thought, one breath at a time.

Recognizing that you need some space or a moment of peace is your first step. Then, when you're ready, move back onto your path, head held high.

Morning affirmation

I am creating a life of peace, ease and comfort.

I am creating a life of intentional happiness.

My life requires neither stress nor struggle.

My life consists of boundaries, clarity and self-care.

I won't allow any negativity into my life.

I am living my life on my own terms.

Ditch your to-do list and write a to-be list

_____

Think about the way you want to be today:

I want to actively listen.

I want to be calm.

I want to be awesome.

Sometimes we focus on just finishing tasks. We forget how we want to be in the world.

Choose one or two qualities you want to embody from your list and move through the day with them in mind.

Be curious

Learning new skills, seeking new knowledge, and discovering new interests are great ways to challenge yourself.

It's not hard. It's just new.

Shift your perspective and nourish your mind.

Meditate daily

A perfect way to find healing and peace is to spend some time each day meditating.

Find a comfortable position that lets your shoulders relax and your hands rest naturally. Pay attention to your breathing – the sensation of air entering and leaving your nostrils.

Accept without judgement whatever thoughts arrive – observe them and let them pass. Be patient with yourself – don't be annoyed if your mind feels restless, just quietly still your thoughts.

This isn't about achieving a specific outcome. It's about being present.

Meditation throughout the day

Meditative thoughts can accompany you throughout the day, wherever you are and whatever you are doing. When you walk. Or eat. Or work. Focus on slow, deep breathing. Notice how your body feels. Observe your surroundings. Listen to the sounds around you. Allow your sense of self-compassion to grow so that you recognize when you are struggling. Be kind to yourself.

A reminder for the restless mind

You do not need to have all the answers right now.

The earth is steady beneath your feet, take comfort in that strength.

You will get there, but right now you are here.

And here is wonderful.

'Finish each day and be done with it ... tomorrow is a new day. You shall begin it serenely and with too high a spirit to be encumbered with your old nonsense.'

———————

Ralph Waldo Emerson

Nothing is permanent

Nothing is everlasting, not your ugly feelings, not the bad days, not the frustrations of a particular situation.

You are not stuck. You have choices.

Make a list of everything that you have found hard today. Then draw a line under it.

The past

_____

You might lie awake wishing you had made different choices. You might regret staying when you should have left. You might remember only the words you left unsaid. You might be angry about the moments you felt you let yourself down.

But it's important to look at your past with gentle eyes.

You did what you did at the time. You can acknowledge that now. Give yourself permission to move on.

Your Body

Physical self-care is as important as water is to a plant. And like a flower, you can bloom and grow if you nurture your body. Your body is speaking, are you listening?

⏳

Sleep

Prioritizing a good sleeping pattern is an act of self-care that has huge benefits for both your physical and mental well-being.

Aiming for seven to eight hours of quality sleep each night isn't just a goal to reach, it's a focus for you to actively establish.

Set yourself regular bedtime routines, have consistent wake-up patterns and make sure your bedroom environment is peaceful. Right temperature. No distractions.

You owe your body and your mind the time to rest.

Bedtime affirmation — letting go of anxiety

I have done my best today.

I let go of what I can't control.

I let go of negative thoughts.

I let go of comparisons.

I focus on my own beautiful journey.

I choose to be calm.

I will have a peaceful sleep.

Bedtime affirmation — calming reminders

———————

I am safe in my room.

I let go of anything that didn't go well today.

My mind, body and soul deserve rest.

I am worthy of healing.

As I sleep, my mind, body and soul will be realigned.

I will wake up refreshed.

Bedtime affirmation – a regular night-time routine

I choose to relax my entire body and sink deeply into sleep.

I give myself permission to rest.

I am grateful for everything that has happened to me today.

I look forward to what tomorrow will bring.

I release the anxieties of the day, they have no power now.

Eat to nourish

Putting the right foods into your body is one of the ultimate forms of physical self-care. A thoughtful diet nurtures your body.

Make conscious decisions about what you eat. Choose healthy foods. Focus on foods which support physical and emotional health. Be strong-minded.

Above all, listen to your body and treat it as your most important possession. It's the only place you have to live.

Jogging can clear your mind

If you don't want to take up a competitive sport as a way of getting physical exercise, going out for a short run is a simple and free way to refocus your mind.

In the beginning, you might only be able to concentrate on moving one foot in front of the other. But as you run more frequently, you'll find that it is a really effective outlet for frustrations. A way to reset your mind. Give yourself some 'me time.'

Short or long, hard or awesome runs – they all carry equal value. You are exercising your body to exercise your mind. And proving to yourself that you are capable of more than you ever thought possible.

'You have to believe in yourself when no one else does – that makes you a winner right there.'

———————

Venus Williams

Prioritize your physical fitness to support your well-being

Sometimes we need a bit of encouragement when it comes to exercising, which is why joining a local sports club or team is a great way of making sure you get an endorphin hit with the support of some familiar faces.

Or you could look out for local gym classes, dance workshops or maybe bootcamp groups. The accountability these provide will make sure you prioritize your physical well-being.

Have fun getting fit

You can increase your heart rate and respect your body without taking exercise too seriously. Turn up the volume on the radio and have a good dance. Take a morning or evening stroll. Go with a friend or listen to a podcast.

Try hula-hooping, skipping or jumping on a trampoline a few times a week. Find a fitness class online and follow along in the comfort of your own home.

Whatever you find fun, turn it up. Twirl around. Bounce along. You'll do wonders for your physical and mental well-being.

Always practise good personal hygiene

Keeping your body clean and in decent condition is an act of self-care.

Showering, brushing your teeth and your hair, and keeping your clothes washed – these all show yourself a kindness and respect. Good hygiene is about giving yourself the confidence and self-esteem that you deserve.

On the days you least want to pay attention to your care, these are the times that it's most important for you to do it. Take a bath. Change your clothes. Treat yourself gently. You're worth the effort.

'Thousands have lived
without love, not one
without water.'

———————

W. H. Auden

Stay hydrated

Water can be one of the most neglected nutrients in your diet, but it is also the most vital.

Keeping yourself hydrated throughout the day is crucial to your body's essential functions. It supports your body to digest food, absorb nutrients, help with temperature regulation, flush out toxins, and much more. It's the driving force of life, so drink up!

Start the day with a glass of water. Keep a bottle to hand throughout the day and sip regularly. End the day with a long, refreshing aqua boost.

'Arise from sleep, old cat,
And with great yawns
And stretchings
Amble out for love.'

Kobayashi Issa

Stretch like a cat

_____

Have you ever watched a cat stretch? One front and one back paw balanced in mirrored perfection, or a spine arched in a state of ecstasy.

Embrace your inner cat and stretch first thing in the morning. Rise up on your toes and reach for the stars.

Follow a yoga or pilates routine, or just be mindful of how you sit, stand and walk during the day. Pull your back straight, with your shoulders low for a perfect posture.

When you stretch, feel a sense of gratitude for your amazing body – from the tips of your toes to the ends of your fingers.

Eat a healthy breakfast

Eating well first thing in the morning will improve your energy levels and concentration throughout the day.

Start the day mindfully with a breakfast that fuels both your body and your mind. Your first meal can kick-start your metabolism, establish a good habit of eating three meals in the day, and improve memory and clarity.

Breathing

It's a simple, everyday intrinsic action that we take for granted. But how often do we breathe properly?

Stand in a comfortable position. Place your hands on your abdomen so you can feel your diaphragm move as you breathe.

Inhale as slowly as you can through your nose. Focus on the image of your stomach being filled with air. Exhale gently through your mouth. Repeat for several minutes.

The transition from chest to abdominal breathing will help eradicate negative energy, bring a sense of lightness, and help your body and mind reset and heal.

3

Your Emotions

Emotional self-care is recognizing that all your emotions play a part in who you are. Understanding and accepting your feelings without judgement is key to experiencing them all.

⌛

A morning mantra

I love who I am becoming.

I realize my worth.

I am worthy of receiving everything I desire.

I am falling in love with myself more each day.

And that feels good — to love the person who has been inside of me this whole time.

'Your new life is going to cost you your old one. It doesn't matter. All you're going to lose is what was built for a person you no longer are.'

Brianna Wiest

Take back your power

Setting boundaries might feel like an unkind or negative action, but they are that peaceful, courageous voice inside that tells you it's time to love yourself.

You are simply saying that you respect yourself too much to let anyone affect you. People will continue to do whatever you allow. It's up to you to decide what's permitted in your life. It's time to take back your power.

Not every day is a great day

Self-care can sometimes mean that you accept you have had a crappy day, but you also acknowledge that these days come to an end. So, you need to:

Get up

Dust off

Yell 'great mother of Maltesers'

Start again

Remember your magic

Every day, when you look in the mirror, tell yourself that you are a sexy little pickle and there is a magic about you that no one else has.

Speak kindly to yourself because you become what your mind tells you.

Practise those self-compassionate words and phrases. Print off some of your favourite sayings to keep by the mirror or next to the front door.

And remember, today is going to be a juggernaut of wonderfulness.

Have you ever seen a serious bird?

A serious sunrise?

A serious flower?

An earnest starry night?

Existence doesn't want you to be solemn.

Existence is a celebration, a little dance.

It's time to join in. Be silly, have fun, enjoy yourself.

Taking a few steps forwards and then one back is frustrating

But perhaps all you are really doing is the cha-cha.

And the cha-cha is a dance of rhythm and fun. So maybe, just maybe, the hiccups along the way of any challenge are life's way of telling you to put on your dancing shoes. Stand up tall. And go with the flow.

Setbacks are frustrating, of course. But you are never going to be right back at the beginning again. Growth doesn't work like that.

Declutter

Sorting out your personal space will bring you a sense of clarity and satisfaction. Shift away from a 'What if I might need it?' mindset and focus instead on letting go.

If you have a big space to declutter, maybe start off small. Work on a bedside table or desk drawer first.

And remember, instead of holding on to items 'just in case', only keep hold of the possessions that bring you joy, are truly useful or hold sentimental value.

Let go of the rest. You'll feel a sense of lightness and liberation in the process.

Ugly emotions are just fleeting negative feelings

They put us in a crappy mood and then make us feel awful afterwards. But's here's the thing, you are not your feelings! Ugly emotions are a part of life, but they don't make you an ugly person.

Next time you are feeling angry, jealous or hostile, try to confront and control the emotion.

Take some deep breaths. Put your hand on your heart. Tell the ugly feeling that you are grateful for its presence, but it is neither welcome nor needed at this moment in time.

Ignite a creative flame

Use your emotions to inspire your artistic side. Jot down your feelings as songs, poetry or musings.

Just as Taylor Swift channels heartache to write some of her more popular hits, you too can use your emotional state to create honest and relatable music. Or try poetry to express raw emotions. If you prefer longer prose, you could keep a journal – this is an effective way of processing your emotions. Make a concerted effort every day to stop and write.

Pay attention to how you are feeling

Try to be attentive to your emotions at any given moment in the day.

Don't hide your feelings from yourself. Instead, acknowledge that they are with you and maybe note down the triggers. Are there certain situations or people that always cause these feelings in you?

Do not fall prey to self-pity when negativity enters your life. Stay strong and write down when and why they came. This can be useful in understanding them better.

'Courage is not the absence of fear, but the triumph over it.'

Nelson Mandela

Inner monologue

Have a positive and constructive talk to yourself when you are faced with a nervous situation or an anxious event.

Challenge the negativity of 'I can't do this.' Focus on a more productive approach: 'I know that this will be out of my comfort zone, but I can and I will do it.'

Take the high road

It can be exhausting to be drawn into arguments with people who thrive in these highly combustible situations. But remind yourself that it's OK to choose the high road, to stick to your own view – your morals – when you feel that's right. Demonstrate integrity and positivity, even if it's not the popular choice.

Choosing a peaceful approach means that you are protecting your values. It doesn't mean you are weak. It means you have the power in every situation – even when provoked or faced with wrongdoing – to know your mind, trust yourself, and simply walk away.

Try to practise self-compassion

Being kind and understanding to yourself during challenging times can be easier said than done. But one way to practise it is to treat yourself with the same gentleness and compassion you would offer a friend.

By sticking to this idea, you will naturally cultivate greater personal resilience. You will find that you respond to challenges with more thoughtfulness, rather than with disapproval and self-criticism.

You can adapt to situations

You don't need to be 'one thing' or 'one way.' You can change as the need arises. This is especially true when you convince yourself that you can't be the way you want to be. If you are the one who always listens in meetings, it doesn't mean you can't present confidently. If you are the person who says 'yes' all the time, it doesn't mean you can't say 'no' on occasion.

You can be soft yet strong,

You can be sensitive but show leadership,

You can be tentative and brave,

You can be many things all at once.

Your Surroundings

Finding a connection with your surroundings, can reset your wellness balance. Spend time outdoors to help reduce stress and feel more contented and peaceful.

⏳

Watching Mother Nature at work is a good reminder that seasons change

Nature doesn't just stop and wait for the bloom of new life in spring. She celebrates and welcomes every change, knowing that each season is an important part of the cycle. And summer, autumn and winter also serve a purpose in the wellness of life.

So, next time you are feeling stuck and unable to move forward, go outside and remember that nothing lasts forever. Every season changes – in life as in nature.

Take a break in nature

Enjoying lunch outside or taking your morning coffee into the garden is a simple and effective way to make you feel good.

Sometimes stepping outside for a short break is all that is needed for a reset – take a stroll at lunchtime or, instead of setting the table at dinner time, grab a picnic blanket and eat outside.

This doesn't have to be in the height of summer either, wrap up warm if it's chilly or fire up the wood burner. It will invigorate your senses and boost your mood.

'Fresh air is as good for
the mind as for the body. Nature
always seems trying to talk
to us as if she had some great
secret to tell. And so she has.'

John Lubbock

Feel the sun on your face

When was the last time you went outside with no purpose other than to feel the sun on your face or the wind in your hair? Take this as your invitation to go outside right now and just soak in the world around you.

Sunlight can boost your mood and help you feel calm and focused, especially during the winter months. Next time you see the sun shining outside, take five minutes to appreciate how it feels on your face. Don't forget suncream!

Make walking a habit

Stroll in the spring air. Jog on a frosty morning. Power-walk on a bright sunny evening. Take in all the energy that the weather and the universe is providing. Go at the same time every day to build it into a routine. You could set yourself a goal of walking for an hour – breaking it up into twenty-minute slots is just as effective as a longer mindful plod.

Want some inspo? Go for a 'colour stroll'. This involves choosing and then looking out for one colour as you walk. Sometimes focusing your mind on a purpose can help reduce stress.

Sparkle stroll

Try going for a walk with a bottle of sparkling water and only return home when you feel a bit more of a spring in your step.

This isn't about pressuring yourself to boost your spirits when you set off for a stroll, but you'll probably find this happens naturally as you go. The deliberate act of saying to yourself 'I'm going out with the purpose of putting a little spark back in my day' could be all that is needed.

Perhaps go on a sparkle stroll with a friend – you could enjoy a catch-up and a confidence-boosting bout of exercise both at the same time.

Bring Mother Nature into your home

There are several ways to feel closer to nature in your own home.

Have fresh flowers or plants inside your house to create a sense of lush, calm serenity and a connection to nature.

Set up a comfy chair or arrange a window seat to give you a place to sit back and watch the world.

Close your eyes and choose a playlist or soundtrack of nature, birds or the sea on your favourite listening device.

Lie still

Take a picnic blanket or soft rug outside and spend some time just lying still, being in the calm of nature.

Peaceful moments allow us to find a sense of relaxation and stillness. They allow the mind to rest whilst making your senses come alive to the sounds of birds and the scents of nature.

Or if you don't want to lie down, take off your shoes and socks instead. Find a small patch of grass or greenery. Feel the earth deep below you, rising up from the soil through your bare skin and filling you with strength and positive energy.

Grow your own

Feel grounded and connected with Mother Nature by growing your own fruit and vegetables.

You don't need an allotment, big garden or a wealth of green-fingered knowledge. You can simply plant and sow seeds from a packet in the garden. Or plant cuttings in a pot.

Many people find that the calming activity of nurturing plants promotes reflection time, peace and a chance to slow down otherwise hectic lifestyles – weeding can't be done at speed!

Swap screens

———————

Instead of reaching for your phone first thing in the morning to check emails, messages or socials, open your curtains and look, really look, at the world outside.

Let the window be the first 'screen' of the day.

Spend a few minutes just looking and watching. Notice the clouds. Are there any birds outside? Can you see any trees? Look around you and quietly observe the wonder of the outdoors.

'A shining star never gets lost in darkness'

Anon

The night sky

Looking up at the stars at night can offer a mindful meditation practice – a healthy way to focus on the present moment.

Gaze up and take in the countless twinkling lights in the night sky to help you appreciate the enormity and wonder of the universe. This can alter your mindset and help you put problems into perspective.

Singing in the rain

Adopt the Norwegian concept of 'friluftsliv': embrace nature regardless of the weather.

Norwegian children are encouraged to go outside all year round, in the driving rain, the icy wind, the fresh snow and the blazing sun. This helps them realize that life continues through ups and downs. It isn't just about tolerating bad days and waiting for good ones. Every moment is precious – rain or shine.

Next time your mind tells you to forget the evening stroll because it's wet, grab your umbrella with enthusiasm, and go and sing in the rain.

Feel the wind

Head to the seaside for the invigorating sea breeze. Or stand on a hillside on a blustery day.

The wind not only blows away the cobwebs, it can help you let go of worries, problems and negative thoughts too.

So, next time the wind is fierce, go outside and focus on what is troubling you the most. Close your eyes and picture the wind rushing through you. Feel the weight of your anxiety lift, and blow far, far away.

Your People

Social self-care involves your connection to others. Your people are the ones who love you and are loved by you. It's important to nurture and protect these relationships.

⧗

'What was done with
love, was done well.'

———————

Vincent van Gogh

Write!

_____

Handwritten letters, cards, postcards or even little notes are such a personal way to send a heartfelt message to a friend or loved one.

These messages show that you have taken the time and effort to send them. And they are perfect keepsakes to look over in months and years to come.

Contagious energy

Some people radiate an energy that is so catchy you can't help but feel alive and sparkly around them.

Celebrate these people. Welcome them into your life every chance you get. Tell them how marvellous the world is with them in it.

That positive energy will pass on to others and then loop back, in a wonderful circle of optimism.

Text an old friend

Been thinking about an old friend? Ping them a text. Miss a former colleague? Write an email to them. Wishing you were in contact with a cousin who lives abroad? Send a voicenote.

These chance, spontaneous messages are such a good way to make a small spark of connection and bring a smile to the recipient.

It's doesn't matter what's in the text, the email, the voicenote. The meaning behind them all is the same: 'I'm thinking about you.'

Ubuntu

Noun: *I am because we are.*

This ancient African word reflects a sense of shared, connected humanity, compassion and harmony. It is a reminder that our lives are interwoven and that kindness towards others shapes who we are.

Think of it like a call to honour – humanity's universal bond. You might not know what others are going through, but you can always show love and kindness.

Work away from home

If you work remotely or from home, it can be easy to isolate yourself in your WFH office space. But don't be afraid to adapt your working pattern once or twice a week to give yourself a bit of company.

Take your laptop to a coffee shop and work among the hustle and bustle of other customers going about their daily lives.

Your interaction might be minimal at first, but in time perhaps you will exchange a few words with the baristas or start chatting to the regulars. It doesn't have to be a deep connection, just some friendly words to break up the day.

Social self-care needn't be planned or structured

Sometimes it's the deliberate act of stopping to chat to a stranger that can leave you feeling refreshed.

Whoever it is – the postman, a shop assistant, someone waiting to cross the road, a fellow dog walker – start with a smile and let the conversation flow.

Hug!

There are hundreds of pieces of research that support the idea that a physical hug is hugely beneficial to your health (as well as feeling pretty good too).

A simple cuddle can reduce stress levels, lower your heart rate and blood pressure, and boost your immune system. This doesn't mean you have to go round snuggling up to random people, just make sure that the next time you see a good friend or loved one, you envelope them in a tight and affectionate bear hug.

'As you grow older, you will discover that you have two hands, one for helping yourself, the other for helping others.'

Audrey Hepburn

Volunteering makes a difference

Charity work is a great way to meet new people and create a sense of shared purpose and community.

Love animals? A local animal rescue will always be looking for help. Enjoy a quiet, calm environment? Why not offer your help at the local library or charity bookshop? Do you like being around children? Volunteer to read with children at a local primary school. Are you an enthusiastic gardener? Go to the local park and see if the gardening group is in need of an extra pair of hands. Love cooking? You could cook or bake for families in need in your community.

Share the sweet treats

Why not use your culinary skills to treat people in your community?

At first, you might want to leave the treats on your neighbour's doorstep with a note. Next time, why not knock on the door and hand them over in person? Or make a batch of cookies and then distribute them round the neighbourhood, enjoying a catch-up along the way.

You may just find that you make their day. A frazzled mum. A retired school teacher. A student studying for exams. Everybody needs good neighbours.

'Trim what drains you, protect what nourishes you.'

———————

Anon

Social media detox

Social media is a place where we build connections – but it should only be with people who bring interest and joy to our lives.

Take fifteen minutes or so to give yourself a social media detox. Unfollow accounts that aren't aligned with your values or outlook on life.

Simple as that. Be as ruthless as you can.

Then refocus your attention onto accounts that make you smile, feel inspired or support your happiness.

'If you are losing your leisure, look out! You may be losing your soul.'

———————

Virginia Woolf

Be a lifelong learner

A great way to bond with others is over a shared interest or passion, which is why starting a hobby or joining a class can be both mentally and socially rewarding.

If you want to stay at home, you could be a virtual participant. Search for different classes that are available online.

If you prefer to join a group in person, keep an eye on flyers, local newspaper articles or social media posts.

Being a lifelong learner gives you a wonderful sense of achievement and belonging.

Don't take it personally

Don't forget – if people upset or hurt you, it's a reflection on them, not on you.

Keep being a dazzling human being, honour your own priorities, and recognize that you have nothing to prove.

Getting to know you is a privilege, remember that.

Your Welfare

Looking after your well-being also means focusing on important life admin - things like spending, earning and saving.

Pause before you purchase

It's so easy to shop spontaneously or at the click of a button. But are these purchases necessary? Or are they perhaps next week's clutter that will soon sit in the back of your cupboard?

Next time you feel tempted to make a snap purchase online, try the 'pause before you purchase' mindset. Add the item to your basket and wait twenty-four hours. If you still want to buy it after you've thought about it, go ahead.

Going shopping at the weekend? Make a list before you go and target just those items. See something else you want to buy? Put it on the list for next time.

Unsubscribe from mailing lists

Sometimes you need to give yourself a friend detox on social media; other times you should do a junk mail clean-up. How helpful are emails that try to make you commit to a product or service that you didn't know you needed? Do offers that guilt-trip you for not taking up their once-in-a-lifetime offer make you feel happy?

If you find it hard to ignore junk mail offering mega promotions, make sure you unsubscribe and stop them hitting your inbox completely. Easy!

'Above all, be the
heroine of your life,
not the victim.'

Nora Ephron

You don't need fashion trends to sparkle

It is so easy to get caught up in the latest craze. After all, everyone has one, so surely it must be good. Or is it? Next time you want to spend your hard-earned money on this sort of product, ask yourself who you are buying it for. Do you really need it? Or are following the crowd? If it's the latter, then give yourself a gentle reminder that you are in control of your own life and you don't need material possessions to shine.

Organize your belongings

Being untidy isn't a problem if you can find your belongings among the clutter. But have you ever bought a new treasure only to discover you already have it?

Try to organize your possessions so that you know exactly what you already own.

Why not have a big declutter? It's very cathartic and afterwards you will feel refreshed and clearheaded.

You could write a list of what you own and keep it in a safe place. That way, you will always have a reference to quickly scan, should the clutter ever creep back in.

People value money in different ways

We all value our money and justify our spending in totally different ways. If you want to get a clear idea of 'worth,' think of the price of an item in terms of work hours.

In the same way that you might consider food intake and exercise in relation to each other, is that pair of shoes really worth three days of work?

If not … well, you know what to do.

A weekly catch-up

Have a regular check-in with yourself, as well as with your partner or friends, about money. It's a good way to align your intentions about spending and saving, share concerns about small worries as well as plan for life goals.

Consider the following: What big expenditures have you got coming up? What are you saving up for? Do you want to change your savings accounts? Can you streamline anything in your life or stop any direct debits?

Be magnanimous

Being generous with money or possessions isn't about giving away more than you can afford, it's about understanding that even tiny acts of kindness can bring great rewards.

You could leave a waiter a well-earned tip; donate your wedding dress to a charity shop; give a friend one of your treasures that you know they really like … There are lots of ways of being kind that will make no more than a small dent in your purse, but will give your heart a huge boost.

'Keep good company, read good books, love good things and cultivate soul and body as faithfully as you can.'

———————————

Louisa May Alcott

Change focus

Where do I see myself in five years' time? The answer to this is all too often focused on items of material value: a bigger house. A smarter car. Designer clothes.

Now change your focus. Stop thinking about possessions and financial success. Consider instead your future health, and see what happens.

For example, 'I see myself ... happy, healthy, with a solid group of friends around me, spending time doing what I enjoy.' Surely that's a much better option for the future!

Believe in yourself

Do you always tell yourself you are rubbish with money? Do you believe that you will never be out of debt? Try to adopt a more optimistic practice. Say or write down positive affirmations. Interrupt the downward cycle. For example:

I work hard and deserve the money I earn.

My finances are in my control.

The money choices I make are in line with my life goals.

A confident mindset will have a hugely positive impact on how you manage your money.

Say 'no' more often

It's amazing the lengths we will go to to please others. Even when it means we will stretch ourselves beyond our financial means. So why do we do it?

Learn to say 'no' more often. It's fair to say 'no' to working overtime with no financial benefit. It's reasonable to say 'no' when you're invited on a hen do that will cost you a lot of money.

Saying 'no' more often is an awesome form of financial self-care. It shows other people your boundaries. More importantly, it shows that you value your time and your money. And these boundaries are to be respected.

Thoughtful gifting

Next time it's a friend's birthday, think carefully about the present you buy them. Wouldn't they love a bunch of flowers fresh from your garden? A hand-drawn card? Home-baked cakes? All of these are heartfelt presents made with a personal touch that won't cost the earth.

Likewise, social media marketplaces are great places to find new gifts at a reasonable price.

And remember, anyone can spend money on a generic gift. You can think of ideas that are not just more sincere and charming, they are cheaper too.

'Her intuition was her favourite superpower.'

———

Anon

Your Spirit

Nurturing your inner self is about
making a true connection with your
soul. Listen to your intuition. Make
time to focus on your spiritual growth.
The world outside is beautiful when
the world within you is peaceful.

⏳

A touch of nonsense

Fancy being a bit goofy? Do it. When life throws you rainy days, jump in the puddle and make no apology for it. Throw snowballs. Splash in the sea. Run wildly after the ice cream van. It's fun and will lift your spirits – and you never need to say sorry for that.

Be grateful for small pleasures

Gratitude doesn't always need to be about the big things. Spot the daily crumbs in your life that bring you real, unadulterated joy. And instead of just ignoring them, actively and mindfully express your thanks. Be grateful for the small pleasures:

The smell of my freshly washed pyjamas.

The way my hair looks today.

How my coffee tastes this morning.

'When you recover or discover something that nourishes your soul and brings joy, care enough about yourself to make room for it in your life.'

———————

Jean Shinoda Bolen

Think about yourself for once

Be selfish and do the things that are just for you. Do this on a regular basis. Don't feel guilty or think you are being egotistical. You are, for once, thinking about your own happiness.

Focus on your inner strength

Have periods of time when you switch off from all the noise. Escape.

We all need a tech detox on occasion and, although that might be hard to do for a whole day, try and factor in small chunks of time when you can mute your phone calls, turn off your texts, ignore your emails. Just be alone with your thoughts. A reminder of your inner strength.

Choose your companion carefully

Who do you want to spend your time with today? Hang out with those who emit enthusiasm and positivity. Not soul-suckers whose pessimism robs your energy.

It might not be a person you choose – animals are a great source of unbridled joy! And because they live in the moment, they remind us to care just about the here and now. Walkies?

Explore your core beliefs and values

Most centres of learning, from nurseries to universities, have an ethos and a set of values that encourage positivity.

If schools are so convinced by these beliefs as a way of encouraging their students, why not adopt their strategy in your own life? Think about your own values. Which beliefs do you value above all others? What is your life ethos?

Take time to consider these principles carefully. Write down your core principles and consider how you can follow those guidelines in your life.

'The world is full of magic things, patiently waiting for our senses to grow sharper.'

———————

W. B. Yeats

Sharpen your senses

Go outside, lie down for five minutes and watch the clouds drift by.

Acknowledge and smile at five pretty things you see on your daily commute.

Savour the scent of flowers.

Admire flower petals.

Stand still. Close your eyes. Listen to the sounds of your world.

Don't be afraid to be seen

Sometimes we want to protect ourselves. We don't want to be seen doing the wrong thing. We want to remain invisible.

This stems from the idea that you need to be perfect.

Don't succumb to this thought.

You don't need to be fearful of an outside voice.
You don't need external validation.
You are you. And that is all you need to be.

Live the wonderful life that is in you!

It can be easy to get so caught up in the day-to-day that you forget to learn, to discover, to seek, to try – to live your wonderful life.

Make it your mission to explore life and experience it as a delicious thirst to quench time and time again.

A compliment jar

Next time someone pays you a compliment, jot it down and keep it in a jar. Watch the pile of notes grow. Enjoy reading them whenever you need a little reminder of what a treasured soul you are.

'Get busy watering your own grass so as not to notice whether it's greener elsewhere.'

———————

Karon Waddell

Water your own grass

Sometimes you might think that all those around you have got enviable, well-organized lives. And you feel like you need to get yourself sorted before you can be happy.

Stop. Breathe. Remember what you have. How lucky you are.

Dig deep into your soul and ask yourself whether you really need to alter anything. Is the true answer a simple one? All you need to change is your perspective.

Treat yourself

Splurge a little. Save up for a small gift for yourself. This is not self-centred or greedy.

It's a way of saying that you value yourself. A little luxury to remind yourself of how important you are – to yourself and to others.

Trust your instinct

Your intuition is your soul whispering a message and hoping that you hear.

It's your gut speaking to you.

If someone makes you feel a certain way, listen to what your body is saying. Trust your instinct. It's rarely wrong.

'If you can't fly then run,
if you can't run then walk,
if you can't walk then crawl,
but whatever you do you have
to keep moving forward.'

———————

Martin Luther King Jr